Our Digital Future

Smart analysis of smart technology

By William Webb

Webb – Our Digital Future

Title ID: 7699713
ISBN-13: 978-1978356177
ISBN-10: 197835617X

© William Webb, 2017

Webb Search Limited, Cambridge, UK: www.webbsearch.co.uk

First published October 2017

Acknowledgements

The insights provided here are invariably the results of interactions with others, listening to presentations at events and reading seminal papers. Many colleagues have stimulated thoughts that have become part of this book and I thank them all.

I sought critical review from a number of individuals and received very helpful feedback from many including Richard Feasey, Richard Harper, Andy Hopper and Geoff Varrall. This does not imply in any way that they agree with all the views expressed here.

I have learnt much from watching my daughters, Katherine and Hannah, use social media and from their views on digitally-related issues. Without that, I suspect this book would miss much of what the next generation is thinking.

And where would we be without those giants of the digital age? Google provided the search tools for finding data while Amazon provided the publication mechanism and sales channel to turn a Word file into an on-demand book.

Table of Contents

Structure of this book 3
1   What's in and what's out 5
    1.1   Setting the scene 5
    1.2   What's in 5
    1.3   What's out 5
    1.4   Can forecasting get it right? 6
    1.5   On data sources, or the lack of them 13
    1.6   What's desirable, what's possible and what's affordable 14
2   Learning from Harry Potter 16
    2.1   Imagination run riot 16
    2.2   The world of Hogwarts 17
    2.3   Star Trek 18
    2.4   Minority Report 19
    2.5   The Hitchhiker's Guide to the Galaxy 19
    2.6   From 1984 to 2001 20
    2.7   Categorising ideas 20
        2.7.1   Here and now 21
        2.7.2   Possible 22
        2.7.3   Implausible 25
    2.8   Summary 25
3   Learning from past predictions 27
    3.1   Introduction 27
    3.2   Famous failures 27
    3.3   Current predictions 29
    3.4   Laws 32
    3.5   Summary 34
4   Key enablers 35
    4.1   Introduction 35
    4.2   The Internet 35
    4.3   Broadband communications and 5G 36
    4.4   IoT 38
    4.5   Virtual and augmented reality 39
    4.6   Artificial intelligence 40
    4.7   Big data 43

4.8	Robotics	44
4.9	Batteries and power	44
4.10	Autonomous vehicles	45
4.11	Autonomous vehicles and 5G	46
4.12	Quantum computing and security	49
4.13	Blockchain	50
4.14	Summary	51
5	The home	53
5.1	Introduction	53
5.2	Home automation	54
5.3	Home entertainment	58
5.4	Family living	59
5.5	IoT and AI	60
5.6	Predictions	61
6	At work	62
6.1	Introduction	62
6.2	The office	62
6.3	Remote working	65
6.4	Non-office working	65
6.4.1	Agriculture	65
6.4.2	Vehicle maintenance	66
6.4.3	Retail	67
6.4.4	Construction	68
6.4.5	Hospitality	69
6.4.6	The digital factory	69
6.5	The changing nature of work	71
6.6	Predictions	71
7	Travelling	73
7.1	Introduction	73
7.2	Planes	73
7.3	Trains	74
7.4	Automobiles	75
7.5	Radical	76
7.6	Predictions	77
8	Leisure	78
8.1	Introduction	78

Table of contents

- 8.2 Cycling as an example 78
- 8.3 Enhanced leisure activities 79
- 8.4 New forms of leisure 80
- 8.5 Predictions 81
- 9 Public services 82
  - 9.1 Introduction 82
  - 9.2 Healthcare 83
  - 9.3 Education 85
  - 9.4 Predictions 86
- 10 Structure of society 88
  - 10.1 An interim summary of our predictions 88
  - 10.2 Society today 89
  - 10.3 A digital backlash? 90
  - 10.4 The impact of society on our predictions 94
- 11 Predictions 95
  - 11.1 Introduction 95
  - 11.2 The world in 2027 95
  - 11.3 The world in 2037 100
  - 11.4 The world in 2047 102
  - 11.5 National differences 103
  - 11.6 Why so pessimistic? 103
  - 11.7 In summary 105
- 12 Structure of the digital industry 107
  - 12.1 Introduction 107
  - 12.2 Our current behemoths 107
  - 12.3 The rise of the Ubers 111
  - 12.4 The digital enablers 112
    - 12.4.1 Connectivity 112
    - 12.4.2 The Internet 113
    - 12.4.3 IoT 114
    - 12.4.4 Devices and software 115
    - 12.4.5 AI 115
    - 12.4.6 Robotics 116
    - 12.4.7 OTT and apps 116
    - 12.4.8 Digital content 117
  - 12.5 The winners and losers 117

13   The future on two pages..118

List of abbreviations

AI	Artificial Intelligence
AR	Augmented Reality
ARPU	Average Revenue Per User
BYOD	Bring Your Own Device
B2B	Business to Business
B2C	Business to Consumer
FTTH	Fibre To The Home
GDPR	General Data Protection Regulation
GPS	Global Positioning System
HVAC	Heating, Ventilation and Air Conditioning
IoT	Internet of Things
MOOC	Massive Open On-line Course
MRI	Magnetic Resonance Imaging
MVNO	Mobile Virtual Network Operator
NFC	Near-Field Communications
NHS	National Health Service
OU	Open University
PDA	Personal Digital Assistant
PVR	Personal Video Recorder
VR	Virtual Reality
V2V	Vehicle-to-Vehicle
W-LAN	Wireless Local Area Network

Structure of this book

There is no set structured method to predicting the future. My approach here is broadly to learn as much as possible from the past, to examine technological progress and identify enablers, to look at a range of different environments and then to build a set of predictions.

Chapter one starts by looking at what is, and what is not included. This is a book looking at the impact of digital on the future so areas such as mobile communications are included but areas such as advances in medical care are not.

Chapters two and three learn from others. Chapter two looks at books on magic and on science fiction to see what their authors imagined in a world where anything might be possible. Chapter three looks at previous predictions to see if there are systematic errors or other biases that I might avoid. Broadly it concludes that until around 50 years ago we tended to under-predict the future but now we have swung the other way and are often much too optimistic.

Chapter four looks at areas of technology where there might be rapid improvements, and in particular those which might enable progress across many areas - as, for example, the iPhone has. It concludes that key future enablers will include the Internet of Things, artificial intelligence and perhaps robotics.

The next five chapters look at different environments such as the home, work, leisure and public services. This provides a structure to consider a wide range of different issues, although there is much overlap between them. The chapters show that in some cases such as the home and public services, change will be slow and minimal, whereas in others such as work and leisure, much more significant changes can be expected.

Chapter 10 asks whether societal issues might have an impact. There is some backlash against some aspects of digital and if this grows it might slow the pace of change. The chapter broadly concludes that although there is disquiet, it will be insufficient to make much difference except in specific cases such as terms of employment.

Chapter 11 then brings it all together with a set of predictions looking 10, 20 and 30 years out. In summary:

- Individuals will see ever-better virtual assistant functionality from their devices as solutions such as Siri improve using emerging AI techniques.
- The home will be broadly unchanged. Some new connected devices such as smart speakers and home IoT products will emerge but home automation will not improve much.
- The office will see widespread deployment of IoT, biometrics and robotics, mostly to save on administrative and maintenance staff.
- Some sectors will make widespread use of IoT to improve productivity such as agriculture and manufacturing. Some, such as retail, will decline further due to changing habits. Some will be broadly unaffected such as construction and hospitality. Vehicle maintenance, which is currently a huge employer, may decline as more electric vehicles are introduced and vehicle sharing becomes widespread.
- Transport will not change materially other than we will be better connected while travelling, have more journey information and see a gradual growth in driverless vehicles (cars, trains, buses, etc).
- Leisure will expand, with each interest area gaining apps, on-line communities, additional functionality and monitoring from IoT devices. This will allow us to spend more time on our favourite pastimes.
- New forms of entertainment based on AI and AR/VR may emerge but will not take up a significant amount of our leisure time.
- Society may become ever-more concerned about the changes wrought by digital, and there may be some push-back.

Finally, Chapter 12 looks at the impact that a world of this form would have on the structure of industry and what this might mean for regulators and Governments. It concludes that the dominant digital companies such as Google and Amazon are very well placed to remain dominant, but that the more traditional suppliers to the digital industry such as Cisco and Ericsson have a less certain future.

1 What's in and what's out

1.1 Setting the scene

When we reflect on changes during our lifetime we think about the Internet, mobile connectivity, the tablet, Amazon, home delivery of groceries and more. We might also reflect on the downside of these developments, high streets ravaged, jobs lost, fake news abounding and the rise of extremist political parties. All of these are associated with "digital" developments – at their heart is the transmission and processing of information in a digital format.

Changes in other areas have been much less noticeable. Cars have display panels now and are somewhat more reliable. Medicine has advanced steadily with DNA analysis, face-replacement surgery and AIDS under control, but treating the common cold still eludes us. Homes are somewhat better insulated, offices more likely to be open plan. All welcome changes, but more of an evolution than a revolution.

Hence the focus of this book on the digital aspects of our future. It is where the action has been in the last few decades and likely where it will be in the next few. I will look at how I expect our digital world to evolve in 10, 20 and 30 years' time - anything more than 30 years is both too speculative and of little value to anyone.

1.2 What's in

Drawing the boundaries around digital is increasingly difficult. Homes are starting to gain digital functionality, as are cars. Digital education might be on the verge of a breakthrough. Communications channels like Twitter are used by Presidential candidates in a way that may have changed the outcome of US, and global, politics. Perhaps better to define what's out.

1.3 What's out

Instead of reflecting on what has changed we might reflect on the biggest issues facing humanity at the moment. Many might respond with climate change, an aging population, the poverty that still afflicts billions, migration, curing cancer or preventing nuclear war. This book has very little to say about any of these directly but a lot indirectly.

Digital could have an impact in many of these areas. Digital monitoring and control could reduce energy usage, reducing emissions. Monitoring could enable the elderly to live at home longer, and robotic carers might even play a role. Big data analytics might help in the battle against cancer, as might throwing more processing power at the problem. But fundamentally most of these are political topics in that they require decisions at national and international level which result in a reallocation of resource. Predicting political direction is near impossible as recent events have made only too clear.

This book does not say much about metrics such as GDP (or even global happiness). This is mostly because these metrics are of little value in a digital world. For example, mobile data usage has grown over 1,000-fold over the last decade but the contribution that mobile makes to GDP has not changed, primarily because the average monthly fee (ARPU, or average revenue per user) has hardly shifted. The same is true of areas such as processing power, home broadband, availability of entertainment and growth of on-line information. Indeed, as an example, Wikipedia adds nothing to GDP but has put Encyclopaedia Britannica out of business, reducing GDP. Better measures are those such as Moore's Law which shows the increase in processing power, but this is of little value until interpreted into the new things that are enabled.

1.4 Can forecasting get it right?

Was it possible 20 years ago to predict the impact of the Internet? Most would say not. But I disagree,

I have made many forecasts over the last 20 years. One of the joys of the Internet is that these can now be readily found and compared with reality. Indeed, I facilitated that process by making quite specific forecasts as to the user experience in given years such as 2015 and 2020. It is entirely reasonable to critically assess those forecasts in forming a judgement as to whether my views here are likely to be well informed. Equally, past success (or failure) is no guarantee of future success (or failure).

What's in and what's out

My first forecast was made in 2000 and published in "The Future of Wireless Communications". This was a time when 3G had yet to be introduced, phones had small LCD displays, Google had yet to receive their key patent and about 99% of homes connected at around 30kbits/s using dial-up. I set out a scenario for 2020 – which was 20 years into the future - which is copied below.

It is the 21st May 2020. John Smith is asleep at home. Today, just as he did 10 years earlier, he has to take a business trip from New York to San Francisco where he is going to make a presentation to a key customer. These trips are now much less frequent as John tends to make more use of video conferencing rather than attending in person, but this time the customer is sufficiently important that he wants to show this by doing things the "old-fashioned way". He will stay the night in San Francisco and fly back the following morning. He notes that much is the same as 10 years before, but some things have improved. Just as before, John booked his flight using his personal communicator interlinked to his company's travel agent. The personal communicator has stored these details and advised John that he will have an early start that day. It has checked expected travel conditions and calculated that John will need to get up at 5.30am, however, early the following morning it will re-check traffic conditions, re-check the flight schedule and determine exactly when John needs to wake up.

At 5.20am the communicator utilises its Bluetooth capabilities to communicate with the in-home network. It pre-warms the water to the shower head and starts to prepare a breakfast for John. It confirms with John's preferred travel site the route that John will take and downloads this to the car, checking that the car has sufficient fuel and that its diagnostic systems have not detected any problems.

At 5.30am the communicator plays wake-up music that John has set as his preference until it hears John say "alarm off". It then presents John with his itinerary for the day. John gets ready to leave the house, drinks his coffee and then climbs into his car. The communicator starts up the car a little before the journey so that the interior is warm, opens the garage door as John climbs into the car and closes the garage door as

John's car turns out of the drive. It locks all the doors in the house and confirms with the house control system that the house is secure.

John still subscribes to a personal news service although he now subscribes directly with a number of news agencies rather than through the cellular operator. His communicator visits these sites and retrieves his daily summaries. For playing back in the car, these are still in audio format, but if John finds any interesting he may say "retrieve video" for the video image to be downloaded and stored on the communicator for subsequent viewing. His preferences are for all sports items to be handled in this fashion, John will view them when he is on the plane. On route, John remembers that he wanted to get the grass cut today. He calls back to the house control unit and instructs it to pass a message to the robot lawnmower to cut the grass today in addition to its weekly schedule. The robot warns him that although it has sufficient fuel for this cut, it will need refuelling before it can complete any subsequent cuts.

Whilst John is conducting these activities his communicator warns him of an important email message. This is in the form of a video clip. The communicator knows that John is unable to view video clips in the car and so plays the audio only. John listens to the audio and dictates an immediate response. Speech recognition has improved much in the last 10 years and John is confident that not only will it recognise his speech correctly, but it will also enhance the syntax and grammar prior to transmission.

Once John reaches the airport, the communicator provides him with a plan of the route to the gate. The communicator uses the W-LAN system to link into the airlines network and electronically check-in, presenting John with his seat number. The communicator will authenticate John as he passes through the boarding gate using Bluetooth. John still queues up early – carry-on baggage space on the aircraft has not improved significantly over the last 10 years.

What's in and what's out

John remembers that 10 years ago he bought his laptop as well as his communicator. Now he sees little point in the laptop. There is no need for a keyboard as speech recognition provides the primary interface, although there are some occasions where John desires a little more privacy. The screen on his communicator is sufficient for the viewing of most images and in any case hotel rooms provide large screens which his communicator can utilise, communicating with them using Bluetooth. He can access any information he desires from almost any location with the high-speed download capabilities of his communicator and the device can store many of his key files on miniaturised storage units. Whilst in the lounge he takes a quick look at his presentation on the communicator screen and makes some small changes.

Although baggage storage has not improved much, flying has become much more pleasant. The seat back has a large display which John can use for many purposes. He can use this as a screen for his PDA, which he does initially to watch his pre-stored news video clips. He can play games based on his PDA such as chess, interactively if he so desires. Communication from plane to ground has become much less expensive so John is happy to receive incoming calls and emails although he still chooses not to make video calls whilst airborne. John can select from a range of films provided by the airways, using his PDA to present him with choices most likely to be of interest initially.

Arriving at the airport, John's communicator informs him that if he leaves now for the customer meeting he will arrive 15 minutes early. Would John prefer to wait at the airport? John asks the organiser for a list of coffee bars at the airport and the prices of a Café Latte. The organiser finds one that meets John's preferences nearest to him, provides him with directions, pre-orders his coffee and debits his electronic purse in-built into the organiser. His coffee is waiting for him when he arrives. Whilst drinking the coffee, John requests a refresher on the history of the company he is visiting and watches the video clip downloaded over the airport W-LAN system. His communicator informs him when it is time to leave and directs him straight to where his hire car is waiting for him.

When John arrives to meet his customer, he talks with him about the changes that have taken place in communications since his last visit. They compare the features and functionality of their latest organisers and some of the latest advances. John's communicator integrates with heart rate and blood pressure monitors interwoven in the fabric of his shirt and monitors the information for irregularities. His colleague has a miniaturised version built into his wrist watch.

In the hotel room, the communicator downloads the room service menu, filters it and presents to John a list of options in the order that it thinks he will prefer. John, however, decides to go out for a meal and asks his communicator to assemble a list of nearby restaurants that have availability in his normal price range. He looks at video clips of the interior of the restaurants and selects his preferred one. A booking is immediately made and the directions downloaded to his communicator - they will be passed to the car when he gets close. Before going out, John makes a number of calls, all video calls, to his home and colleagues. He then looks at his videomail messages and replies to some.

On his return to the hotel room his communicator dims the lighting, locks the door and prepares the room for John to go to bed. It informs John of his schedule the following morning and suggests an appropriate wake-up time. John accepts. Before going to sleep he reads some more of a recent novel displayed on the screen of his communicator.

John has now achieved his vision of being able to talk to any person or machine in any manner that he liked. His personal communicator is now controlling much of his life including the environment around him as he travels and filtering the information that he is presented with. The communicator has become truly indispensable in helping John organise his life.

What's in and what's out

Recall that this is for 2020 - so there are still a few years left to realise those things not yet available. Broadly, though, it appears to me to be close to reality, albeit an over-prediction in some cases such as home automation. It suggests that forecasting 20 years out with sufficient accuracy to be useful is possible.

My latest predictions were in "Being Mobile" written in 2010. This was a somewhat different book that looked at technical advances being discussed at the time and key vertical segments and services that might be delivered. Its conclusions were:

> We started this section with a look at users and how their behaviour has changed. One of the key messages was that the industry frequently mis-forecast what users wanted, often delivering services that were unsuccessful such as video calling and being surprised by the success in other areas such as ring-tones and texting. The chapter concluded that the mobile phone would become ever more personal, used for social communications, entertainment and distraction much more than for "valuable" communications. This would continue to be difficult for the industry to understand with the result that value would often be captured by others such as Apple, social networking websites or publishers.
>
> Sensor networks offer solutions to a number of problems. They can monitor health, better control buildings to reduce energy consumption or monitor the environment more widely. The ideal of tiny sensors which can be "scattered" and then forgotten seems unlikely as the problems of powering them are just too great. However, sensors in building and worn by individuals are much more likely and indeed already emerging. These can make use of a number of developed standards and of a range of network architectures depending on their requirements. Sensors will require a mix of small home cells and / or mesh networks to interconnect them.
>
> Home cells will be a core part of a hierarchical network architecture. This will become increasingly critical as data volumes rise on cellular networks. As this emerges devices will become "always best

connected" although operator business models may act to prevent this in the short to medium term. In the longer term we can expect operators to adapt to the hierarchical model.

Healthcare is an obvious area where wireless can bring benefits. Sensors can check on the condition of individuals and sound alerts, making it possible for some to live longer in their home. Wireless in the hospital can bring efficiency benefits. Wireless healthcare broadly builds on sensors, mesh networks and small cells in the home or hospital – all of which are available today. However, the pace of change in healthcare is slow due to a number of reasons including funding issues, risk aversion and the difficulty in developing standards.

The prognosis for transport is similar to that for healthcare. Wireless can bring benefits including better route guidance, improved vehicle safety by avoiding collisions and sharing information, luggage tagging at airports and passenger communications from planes and ships. However, any solution that requires collaboration between manufacturers, governments and others is difficult and slow. Such solutions are likely to be overtaken by standalone systems that can be fitted (or retro-fitted) without any need for widespread agreement even if the functionality is not quite as good.

Entertainment has been one of the largest users of wireless communications and this will continue to be the case. The biggest demand comes from video content and a key uncertainty in the future is whether users will predominantly "side-load" this content from a home server to their device or will stream content to their device when out of the home. In practice, the costs of the latter suggest it will likely only be adopted for material which has a high value when live (eg news, sports events) or when the device is in a hotspot and so able to download at a much lower cost.

Assisted living services are critical for an aging society and wireless can play a role. Much of this is the same as for healthcare – sensor networks coupled with the provision of information can allow people to live in their home for longer. There may also be a role in providing engagement services to allow people to remain active members of society even if they cannot get out so much and allowing some to telework so they can continue to contribute. Like healthcare and transport, the issues here are not so much a lack of technology but those of funding, conservatism and the difficulty in agreeing standards.

Overall the picture is one of wireless having adequate technology but being held back by the complexities of introducing new services into environments such as transport and healthcare. Where little agreement is needed rapid advances can be made – the iPhone is a classic example of this. Where agreement of manufacturers and government bodies is needed progress is typically very slow – so much so that the optimal solutions may be side-stepped by less optimal but simpler to implement alternatives.

The approach in this book reflects most closely this last set of predictions in segregating by environment. It will build on the conclusions there about the difficulties of introducing services in some areas, which have proven to be just as problematic as imagined.

Overall, I would argue that my predictions as to services delivered, end-user device capability, speed of adoption and deployment of new technologies and concepts have been pretty good in such a fast-changing world. It is possible to make predictions that have a good likelihood of being accurate.

1.5 On data sources, or the lack of them

We live in a world that values rational, evidence-based analysis. Many might be more comfortable with a book that sets out previous trends and uses them as an "evidential basis" for the future. I have broadly not done this.

My first issue is that evidential analysis can only be backward-looking. To count as evidence, something must have happened, and if so it is in the past, not the future. Where change is slow and predictable then this approach can work – for example understanding future population growth would clearly use evidence from the past. But where change is rapid then previous evidence is unavailable or even misleading. What evidence, for example, was there for the mobile data explosion that occurred in 2007? Or for the rise of the autonomous car when looking from 2010?

My second issue is that much of the evidence that might, perhaps, have value, does not seem readily available. For example, I would like to know the total sales of robotic vacuum cleaners over the last decade as I suspect it would reveal limited growth. I have not been able to find such evidence. This may be partly due to my incompetence, but I suspect more due to the reluctance of manufacturers to reveal such data if it might cast doubt on their growth story. A lack of evidence may be evidence of a lack of success. Or it might just be that nobody has published it.

My third issue is around using forecasts from others. Forecasts often exhibit the "hockey stick" of growth; typically occurring two-three years hence (any less is not credible, any more is beyond the funding window for investors). Such forecasts get moved a year to the right each year until either everyone gives up, or the long-anticipated growth finally materialises. IoT currently falls somewhere between these two outcomes. My point is that such forecasts are rarely reliable and I prefer to use my own intuition rather than build on predictions that others have made which may contain many intended and unintended biases.

Fourthly, as you will see, I suggest key trends include AI and robotics. I struggle to see what evidence there is for AI that can be charted and used for extrapolation. Computer processing power and memory can be predicted based on prior evidence, and these are important enablers for AI, but creativity and breakthroughs are still needed. These things cannot be evidenced.

As a result, there are not many charts or data sources in this book. Instead, I rely more on logic and process, as described below.

1.6 What's desirable, what's possible and what's affordable

The process that I am going to follow is firstly to ask what we would like in an ideal world. To do this I turn to magic and science fiction – their creators often endow such worlds with wondrous capabilities that we would love to have, such as teleportation.

Next, I look at what is technically possible. Sadly, teleportation is not, but some of the smart homes envisaged, for example, most definitely are. To understand this better I discuss some key trends such as data rates, battery life and other underlying enablers needed to make most visions happen.

However, technically possible is not a sufficient hurdle. Flying cars are technically possible but rarely seen. New products and services need to be affordable. More specifically, users need to decide that they would rather spend limited cash on them than the myriad of other choices available. Smart homes have tended to fail at this point, for example.

From this background, I then look at specific environments or topic areas such as the home, office, transportation, education, work and leisure. This segmentation is mostly to provide a framework to the discussion and there will, of course, be overlaps between these areas.

2 Learning from Harry Potter

2.1 Imagination run riot

Henry Ford is widely quoted on the subject of a lack of imagination – that if asked about the future of transportation most would have wanted a faster horse. Mostly, our world does progress by evolution and indeed, at present most are asking for a faster mobile phone. Sometimes, faster, or otherwise better, is all that there is. But on rare occasions an entirely new service, product or concept appears. The iPhone was a good example. Predicting such things is extremely tough, but some have tried, predominantly the authors of books in the genres of science fiction and of magic. In these worlds, there are no constraints, or the constraints differ radically from those we face. Imagination runs riot. And sometimes, the real world follows along behind – that Star Trek communicator looks uncannily similar to the Motorola StarTac phone of the late 1990s.

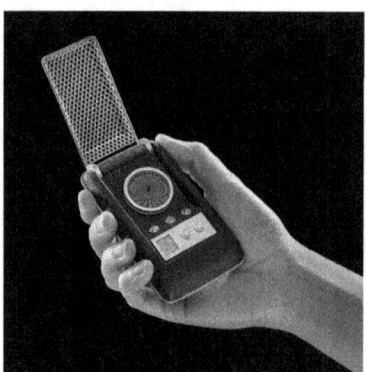 

Figure 2-1 - The Star Trek communicator and Motorola StarTac of the 1990s

In this chapter I take a look at the worlds created by some of the most influential of the science fiction and magic books of the past 50 years or so. I have not gone back further (for example to Jules Verne) because it is clearly harder to get predictions right the further away you are from the prediction date. I have mostly avoided books that discuss dystopian societies since they do not spend so much time imagining what we might want as what might go wrong (but I make an exception for 1984). My selection is rather random, based on those books that

I've found most useful or that are most widely quoted when the future does arise (Star Trek seems to play a major role here). I'm no Trekkie though, so likely have missed a number of wonderful ideas. That is not critical, the aim of this chapter is to brainstorm, to cast a wide net and see what I catch.

2.2 The world of Hogwarts

There are so many wondrous inventions in the Harry Potter books it's hard to know where to start. Wands and broomsticks, of course. My favourite is the Weasley's clock which looks like a grandfather clock but where the hands show the location of each family member, with the 12 o'clock position reserved for "in mortal peril". Other key inventions include the marauders map which shows the location of everyone in the building or vicinity and, of course, the invisibility cloak – a way to become invisible under a piece of clothing. Indeed, invisibility features in most magical stories such as the use of the ring in the Lord of the Rings. Time travel also makes an appearance with the time-turner which enables short-term time travel (allowing Hermione to attend multiple simultaneous lessons).

There are many inventions that serve to help Harry and friends do battle with the dark forces. These include the sneakoscope – a detector for dark arts, the secrecy sensor which vibrates when it detects lies, and foe-glass – a mirror that shows how close enemies are.

Transportation features heavily, nearly always involving flight. Naturally broomsticks, but also the flying Ford Anglia. Like many science fiction settings there is also teleportation in the form of floo powder and fireplaces.

Then there are inventions that just enliven the everyday such as the howler – a letter that opens itself and shouts out its contents to the embarrassment of the recipient. The deluminator can put out lights from a distance. The remembrall glows when the owner has forgotten something (but unfortunately does not tell them what it is they have forgotten) and in a similar vein the pensieve allows memories to be stored then reviewed in a virtual-reality type manner. Wizard chess enlivens the board game with animated pieces which drag others off the board. Hermione's handbag is Tardis-like in having near-infinite capacity. The quick quotes quill takes notes by itself – albeit giving them its own salacious twist.

2.3   Star Trek

Star Trek is often seen as the place that the future is premiered. Perhaps this is because it has been running for so long (since 1966), because it is so widely seen, and because the venue of a spaceship allows for unconstrained imagination. Much that passes for a new product can be seen on prior episodes of Star Trek – for example something akin to the Apple AirPod earphones were worn by crew members over 40 years ago.

Robotics features in a number of places, often called androids. These creatures are indestructible, have super-human powers but are also good friends.

There are many features connected with the space ship, the most famous of which is warp-drive, a faster-than-light propulsion system needed to get the spaceship to other planets in a sensible timeframe. Star Trek had its own invisibility in the cloaking device, the difference with Harry Potter being this could hide the whole of the Starship Enterprise rather than just individuals. In a similar vein, force fields could be set up around the ship or around areas, protecting the occupants inside. Inside the ship the holodeck provided holographic imagery which could be used for communications or entertainment.

For entertainment they had "The Game" – a virtual reality type of activity that also involved direct stimulation of the brain's pleasure centre. It was hopelessly addictive. They had tablets long before the iPad or even its predecessor, the Apple Newton.

Another whole set of inventions were medically-related. Nanoprobes entered the human body and could repair damaged cells, but also had a darker side in that they could take over functions and control the human. The dermal regenerator could instantly heal cuts and burns. Hypospray got medicine into the bloodstream without needing an injection using some form of high-pressure jet. The medical scanners, tricorder and biobeds could provide instant diagnostic information without needing to actually touch a patient.

For weaponry they had the phaser – a gun-like device that could stun or kill and that could deliver huge bursts of energy to melt rocks and so on. When encountering other races they had real-time translation. This is another widely-used device in science fiction.

A key device was the replicator – a kind of 3D printer that could create replicas of whatever was fed into it including food and clothing. It also had the ability to then recycle whatever it had made back into constituent parts ready for its next task.

Most of all, Star Trek will be remembered for the transporter ("beam me up Scotty"). Invented more as a way to save money on special effects by avoiding the need to have the ship dock, this took apart the person or thing placed at one end and then perfectly recreated it at the other. Like floo powder it provided instant transportation.

2.4   Minority Report

Minority Report is often quoted, more for the ethical questions it raises about arresting someone before they actually commit a crime. But its futuristic wall-to-wall transparent video displays were hugely memorable. When Microsoft launched a multi-sensory surface called Microsoft PixelSense they promised it "will feel like Minority Report". Another technology now available is retinal scanners, although they are not yet as slick as those in the film. Similarly, personalised advertising is coming along rapidly and one day will likely provide out-of-home tailored messaging in the way that happens in the film.

Insect robots made a few appearances, often being used for reconnaissance missions.

2.5   The Hitchhiker's Guide to the Galaxy

The Hitchhiker's Guide was more satire than serious science fiction. The Guide itself was akin to a large Blackberry, able to provide encyclopaedia knowledge on everything. Like Star Trek it had its teleportation in the form of matter transference beams which required immediate eating of peanuts afterwards to stay alive. It had a replicator in the Nutrimatic Drinks Dispenser but unlike Star

Trek this one failed to do anything but deliver drinks that were "almost, but not quite, entirely unlike tea".

The Joo Janta 200 Super-Chromatic Peril Sensitive Sunglasses were designed to help the wearer develop a relaxed attitude to danger by turning completely black at the first hint of trouble. Another wearable, the thinking cap, appeared to enhance brain function, powered by a lemon.

One well-remembered invention is the babel fish, another variant of the instant translator. This was a small fish placed in the ear that performed the translation function.

2.6 From 1984 to 2001

Both 1984 and 2001 hugely over-predicted the changes that would occur by the time the dates of their titles arrived. Both were also more dystopian than utopian.

1984 is famous for its surveillance, using a TV screen that could also monitor the occupant. Other ideas now look quaint, such as the tubes for sending documents around. However, we have yet to realise the art-creating and novel writing machines that were envisaged, albeit automatic news stories can be generated from various journalistic feeds.

2001 was more about space travel but did introduce the idea of artificial intelligence run amok in the form of HAL[1], the on-board computer that clearly has a mind of its own.

2.7 Categorising ideas

I could keep going, but it is clear that there is a fair degree of overlap. In this section, the key ideas relevant to this book are separated into three categories (so things like warp-drive are not included!): those available now, those that are technically possible but not commercially available, and those that are likely completely impossible, at least within the timeframes considered here.

---

[1] Allegedly, the abbreviation was a play on IBM, the letters being one prior in the alphabet.

This categorisation is necessarily subjective. Some might argue that the full form factor envisaged by the author is not available, or that variants of the concept are. It likely does not matter overly; the aim is to capture a list of ideas that would be great in principle that I can explore further later on.

2.7.1   Here and now

Location tracking, as shown in the Weasley Clock, is entirely possible[2]. Various apps allow you to track friends nearby or to find lost phones. Sadly, it is not possible to buy the clock itself, but that is because the cost would be high for a service that is already available on phones and tablets.

Helping with memory, along the lines of the remembrall, is also possible. Various approaches from appointment reminders, to-do lists, and smart assistant commands such as "remind me when I pass a florist", suffice for most of us. There is certainly no longer any excuse for forgetting birthdays!

Medical scanners as seen in Star Trek are also widespread in the form of MRI scanners. Automatic diagnosis using AI is not quite here yet but clearly on route.

Until recently, real-time translation (the babel fish) looked to be something for the future. But the advances in Google's translate functionality and voice recognition are astounding – a good example of the use of machine learning to rapidly gain competence. It clearly will not be long before this becomes a viable way to translate (and means human translators will increasingly be out of a job), although translated material often has a certain dissonance about it, lacking a human touch for style and beauty.

Large surfaces that provide touch-screen capability and even gesture recognition are also now available using technology such as that pioneered in the X-Box. That they are not widespread is more a function of the cost and the need for a large room to house them, and that the benefit they bring is relatively small.

---

[2] Microsoft demonstrated a trial of this, see https://www.microsoft.com/en-us/research/publication/the-whereabouts-clock-early-testing-of-a-situated-awareness-device/. It showed some interesting findings as to the social value it could provide.

2.7.2 Possible

While location is possible, the marauders map cannot be realised in its full glory because foldable displays are not yet available. This is an area where there has been much interest, and demonstrations shown over the last 20 years, but the closest we have got to implementation is the curved screen in some Samsung phones[3]. Rollable and foldable is expensive and fragile. Such devices would need a lot of very careful treatment in everyday use without the hard outer casing that protects phones and tablets. The benefits they bring would be very limited. So while possible, foldable displays seem very unlikely to be implemented.

The pensieve does not look too far off. A mix of video recording and VR processing should enable events to be captured and made near-real for those reviewing them.

Flying cars and jetpacks are also clearly possible and have often been demonstrated. But they have never made it out of the prototype stage – as epitomised in Peter Thiel's oft-repeated quote that we wanted flying cars and instead got 140 characters[4]. My view is that flying transport will never become personal (in that the average person will never own a flying car, or similar). If we did, the benefits of faster travel would soon disappear as congestion shifted to the skies (and to landing pads). Flying cars would need to be hugely more reliable than their earth-bound cousins and would burn vastly more fuel. Accidents would almost always be fatal. The slim benefits are never going to outweigh the costs. So no more discussion about flying cars here.

---

[3] In passing, it is worth noting that in the early 2000's a number of companies were demonstrating rollable displays and as a result many predicted the demise of the phone and the rise of all sort of new form-factors. In practice, none of this has happened. This is a good example of how easy it is to be influenced by concept devices and assume they herald the future.

[4] Peter Thiel has lectured that the rate of technological innovation is decelerating despite our collective belief that there are smart scientists in labs somewhere, working to solve our problems. He believes Government regulation is largely to blame, and that inventive minds are too focused on the internet and too neglectful of the world of things. Instead of progress, real food and energy prices are higher than they were forty years ago, and flying from L.A. to San Francisco has actually got slower.

Androids and robots are also possible. Robots are widespread on production lines and increasingly used in large warehouses and similar. Humanoid robots that can interact with, for example, elderly patients are steadily gaining ground to the point where some patients are happy to treat them as a surrogate companion. But it is when robots need to move across anything other than a smooth flat surface that the problems start. We have had robotic vacuum cleaners and lawn mowers for over a decade, but they have made very little progress. Vacuum cleaners cannot climb stairs, open doors or move obstacles out of the way. They also tend to be relatively under-powered compared to a conventional vacuum cleaner. Robotic lawn mowers cannot open the gate to get to the front garden or even open the shed to get out. They are also underpowered and can struggle with thick grass.

Robotic mobility is not a problem going away any time soon. Overcoming it requires complex mechanical mechanisms that make the robot much more expensive as well as needing maintenance. Robot intelligence will grow, robots in manufacturing will become ever-more present, but the robotic house-attendant still looks unlikely to become mainstream. There may be exceptions in some cases such as those living at home who are disabled or infirm where the second-best and more expensive option makes sense.

This conclusion might seem rather pessimistic - some expect robots to rule the world, I'm doubting whether they can even climb the stairs. But recall that the history of robots is long, and that there are no significant improvements in mechanical parts foreseen.

Micro-robots, such as those in the bloodstream or swarming around on reconnaissance duty, are, to some degree simpler. The mobility problem often goes away, or can be solved with flight. The new problem is miniaturisation, especially of mechanical components and power supplies. Robustness to survive harsh environments is also a challenge, as are unexpected challenges from small size, such as the huge impact the surface tension of liquid surfaces can have on very small devices[5]. Significant progress in the next 10-20 years seems possible, but whether such devices will materially change our lives is unclear.

---

[5] See, for example, the work of Richard Jones, http://www.softmachines.org/wordpress/

Phasers appear to be a mix of tasers and lasers, both of which are widespread. We cannot melt rock with a handheld device but we can stun at a distance. However, the benefits of such devices would only accrue to police and military (hopefully).

The replicator is a kind of 3D printer which can analyse items presented to it. We have good 3D printing capability and image recognition of simple objects is possible. A device that can send an image back to the Internet for recognition and where the Internet can then provide the code for replication seems plausible. However, the range of materials that can be 3D printed is limited - plastic is easy, fabric harder. 3D printers can produce metallic parts, but such printers are very expensive and very large. Generic replicator functionality would seem well beyond the economically possible and practically viable for the foreseeable future. But a device that can be presented with a simple part and which can automatically fabricate a replacement in plastic would seem very viable and likely to appear in the next decade.

The Holodeck – holographic displays - is another long-standing area of research. Many displays have been demonstrated, the best often based around a rapidly rotating vane onto which an image is projected. This needs to be encased in a glass dome. The effect is quite good, but a long way from the magical 3D appearance of a person in science fiction. In practice, there cannot be an image without something to project it onto. Holographic displays will always be curiosities rather than mainstream home entertainment engines.

The surveillance screen from 1984 is little more than a smart-TV with in-built camera. That is clearly possible, but not something many would want.

The AI exhibited by HAL is maturing fast. Huge advances have been made in recent years in linguistics, machine learning and in using the Internet to train programs. High-profile victories at games such as Chess and Go have fuelled the view that machine intelligence is rapidly surpassing that of humans. Whether AI can move from doing very well at well-specified tasks to being good at something more general remains to be seen, and will be discussed in coming chapters.

2.7.3 Implausible

Invisibility is a tough challenge. I did see a demonstration of a coat comprised of millions of LEDs woven into the fabric. A forward-facing camera sent an image to the back of the coat and vice versa. This did give a bit of camouflage but while you could not see the person (except their head) there was clearly something odd going on. Real invisibility is too difficult.

Hermione's handbag is a wonderful idea but breaches the laws of physics. Something cannot be bigger on the inside than the outside.

Teleporting, either with or without floo powder is also not possible. There is no way to instantly send matter from one point to another. And time travelling is even more unlikely. That is a shame because the world would be a more interesting place if these things could be arranged.

2.8 Summary

A first observation is that we have achieved much that was science fiction not too long ago. The personal communicator was wishful thinking 40 years ago and the language translator still apparently out of reach only five years ago. Even for films as recent as Minority Report, we have delivered most of what formed a futuristic landscape (except, happily, the pre-cognitisation of crimes about to happen). This is a hugely important point - over the last 20 years or so we have realised much of what was assumed to be an impossible utopia. Indeed, so much so that there is little left to imagine that is realistically possible. That, alone, suggests that the next 20 years might not be such a period of dramatic change as the last 20.

Areas where there is promise of improvements include:

- Even better language translation.
- Robots with more intelligence.
- Micro-robots - but it is not clear what they would be used for.
- Replicator-3D printers.
- Artificial Intelligence (AI).

Would any of these make a real difference to our lives? AI is perhaps the most interesting as it is a general-purpose advance somewhat akin to the Internet. Intelligent robots are related in that their intelligence will probably come from AI, the only difference is that it will be encased in a humanoid, or similar, form-factor.

I suspect what you might have found most surprising is my views on the limitations of robots that need to move, my utter scepticism on flying cars and the unlikely emergence of foldable, rollable or holographic displays. In all these cases, it is not the technology so much as the economics that I see as the issue. These things are expensive and unless they bring very substantial benefit will not be embraced in large number.

Having learnt from the creative talents of authors or science fiction, I now turn to look at what we can learn from predictions previously made. There are plenty of very inaccurate predictions, is there anything that can be done to avoid this?

3 Learning from past predictions

3.1 Introduction

Bill Gates must rue the day that he said that 640k ought to be enough memory for anyone. We seem to have settled out at roughly 10,000 times that amount, so while he was right that there was an upper limit that enabled most requirements, he was wildly wrong about where that limit lay.

As I showed in Section 1.4, my track record is mostly good. This chapter looks at the lessons that can be learnt from predicting the future and whether my predictions can remain accurate.

3.2 Famous failures

There are many lists on the Internet of famously bad predictions[6]. Here is a representative sample:

- "There is not the slightest indication that nuclear energy will ever be obtainable. It would mean that the atom would have to be shattered at will." - Albert Einstein, 1932
- "This 'telephone' has too many shortcomings to be seriously considered as a means of communication. The device is inherently of no value to us." - Western Union internal memo, 1876
- "Rail travel at high speed is not possible because passengers, unable to breathe, would die of asphyxia." - Dr. Dionysius Lardner, 1830
- "I think there is a world market for maybe five computers." - Thomas Watson, chairman of IBM, 1943
- "X-rays will prove to be a hoax." - Lord Kelvin, President of the Royal Society, 1883
- "Everyone acquainted with the subject will recognize it as a conspicuous failure." - -Henry Morton, president of the Stevens Institute of Technology, on Edison's light bulb, 1880

---

[6] This one is adapted from http://list25.com/25-famous-predictions-that-were-proven-to-be-horribly-wrong/

- "The horse is here to stay but the automobile is only a novelty—a fad." --The president of the Michigan Savings Bank advising Henry Ford's lawyer not to invest in the Ford Motor Co., 1903
- "Television won't last because people will soon get tired of staring at a plywood box every night." --Darryl Zanuck, movie producer, 20th Century Fox, 1946
- "There is no reason for any individual to have a computer in his home." --Ken Olson, president, chairman and founder of Digital Equipment Corporation, in a talk given to a 1977 World Future Society meeting in Boston
- "The wireless music box has no imaginable commercial value. Who would pay for a message sent to no one in particular?" --Associates of David Sarnoff responding to the latter's call for investment in the radio in 1921
- "There will never be a bigger plane built." -- A Boeing engineer, after the first flight of the 247, a twin-engine plane that holds ten people
- "I must confess that my imagination refuses to see any sort of submarine doing anything but suffocating its crew and floundering at sea." — HG Wells, British novelist, in 1901
- "The world potential market for copying machines is 5000 at most." — IBM, to the eventual founders of Xerox, saying the photocopier had no market large enough to justify production, 1959
- A rocket will never be able to leave the Earth's atmosphere." — New York Times, 1936

Interestingly, almost all the major failures are under-predictions. They are about ideas that appeared crazy at the time, but became mainstream, or about product innovation that allowed a device to become cheaper, smaller, more flexible and hence much more widely deployed. They amuse us because we imagine that those in powerful positions ought to be good at predicting the future. After all, who could be better placed to comment on computer memory than Bill Gates?

There are two points to make here, that new ideas often initially appear inferior to the existing solutions (a form of innovator's dilemma) and that those who have made a successful career around a particular industry are often the last to spot change.

The first point is explained well in the book - "The Innovator's Dilemma"[7]. In many cases, when a new idea is first suggested it is ill-formed and embryonic. Compared to existing products that have been honed over the years it often looks amateurish. But as it evolves it can side-line the incumbent. Classic examples include small hard disk drives, digital cameras and so much more. The lesson here is not to compare the new idea directly with the incumbent but to either imagine what the new idea might become or to compare it with the incumbent in its initial instantiation.

The second point is that those deeply involved in a company or product can often find it hard to step away from the daily management and think about what might be. They will be naturally disinclined to think about things that would disadvantage them. We might think these sorts of people should know better but often they are some of the worst predictors of the future.

Finally, many of these predictions are now over 50 years old. We have become much less likely to under-predict as we have got used to a rapidly changing world. Instead, over-prediction tends to be the problem - the assumption that we will have flying cars by 2010. In my case, prediction errors tend to be over-prediction. It has been said that we tend to over-predict the short term and under-predict the long term. In "The Road Ahead" Bill Gates (perhaps trying to make up for his earlier errors) said "We always overestimate the change that will occur in the next two years and underestimate the change that will occur in the next ten." Perhaps in reaction, forecasters seen to have become more radical about the longer term than ever. I examine this in the next section.

3.3 Current predictions

The Internet is full of predictions. After all, anyone can make them and it is fun to do so. Here, for example, I discuss a set of predictions published in The Telegraph[8] in 2015 for the year 2045, but many other sources would be equally good.

---

[7] For more details see https://en.wikipedia.org/wiki/The_Innovator%27s_Dilemma
[8] See http://www.telegraph.co.uk/technology/news/11943575/Back-to-the-Future-Day-Five-experts-predict-life-in-2045.html to read the full predictions.

The first prediction, from Alex Ayad, head of Imperial College London's Tech Foresight Practice, looks likely to be an over-prediction. Alex has three main ideas: direct brain stimulation, living material for city centre buildings and invisibility cloaks for buildings. His first idea is that we can purchase emotions on-line, perhaps to go with particular pieces of art or experiences that our friends have had. These can then be implanted into our brains for a more authentic experience. The second idea is that buildings will be made of living materials that adapt and mould to the environment, although there are no details on what these might be. The final idea is based on some mathematical theory that some artificial materials might just be able to bend light and extrapolating this to invisibility cloaking of ugly features and buildings. These are all far from proven, and even if technically possible do not seem to be things we would pay much for. This seems to be a case of Alex having seen something mentioned as a research topic somewhere and extrapolated that into mainstream product in 30 years (which is not a long time in the scheme of, for example, city centre buildings).

Oren Etzioni, chief executive of the Allen Institute for Artificial Intelligence, suggests that AI will solve some of our biggest questions by 2045. He gives the example that AI could scan millions of medical articles to extract hints about possible cures for cancer. But he does not believe that AI will have reached human intelligence levels by 2045. Instead he thinks AI could work with us, helping us to find solutions to problems that have so far proved intractable. This seems to be much better grounded. He extrapolates forward from today's AI to one that is similar, but more accomplished and shows how real value could be added.

Tamar Kasriel, founder and MD of Futureal, future-focused strategy consultancy, seems to be reading the same material as Alex. He covers three areas. His first is VR which he thinks will lead to a range of experiences indistinguishable from the real thing, such as hoverboarding. His second prediction is "upgradable humans" with better eyesight and hearing and a range of prosthetic add-ons. Finally, he suggests buildings will be self-powering with solar panels built into their outer materials. The first and third sound plausible - VR experiences might well be life-like by then, and solar panels are already available within windows and roofing materials. However, blending humans with artificial body parts, unless necessary after amputation, seems less likely given the risk aversion in the medical world.

Richard Watson, futurist, writer and founder of online magazine What's Next, is more pragmatic and picks up on many of the ideas that arose in the previous chapter. He is also interested in emotions but in the more realistic way of a computer reading someone's emotions through signs such as facial expression and heart rate, rather than trying to insert new emotions directly into a brain. The computer might then decide, for example, to hold back the delivery of emails if it detects the person is stressed, or for cars to adapt to improve safety if they feel the driver is angry. Emotion detection has already been demonstrated through facial recognition so this is all very plausible. He believes that by 2045 we will have swarms of insect-sized robots which might be used for crop pollination or testing air quality. This was mentioned in the previous chapter and seems possible, although the value even in these applications is unclear. Finally, he suggests 3D printers could print food such as a pizza. This seems far-fetched to me - getting the ingredients would be difficult, the cost would be high and I suspect the taste not as good as the "real thing". When food deliveries from supermarkets are quick and convenient this seems a step backwards.

Peter Cochrane, OBE, advisor and former BT chief technology officer, who I occasionally work with, is similarly sensible, building on prototype devices that already exist and trends towards connected homes that are already happening. He talks about bathrooms where mirrors, scales and toilets can check on bodily functions, analysing saliva, skin salinity. These will then be processed and sent to the kitchen where appropriate smoothies will have been prepared and menus suggested. This is also all possible, although I am unsure whether many of us would value this sufficiently to pay for it, a point I return to when I discuss the home environment in Chapter 5.

Mark Drapeau, head of content, World Future Society and editor, The Futurist picks up more on politics and business, foreseeing improved productivity and reduced production costs enabling social good. As many forecasters do, he talks up the current darling of innovation - in this case Uber - and assumes it will still be leading the way in 30 years' time. He suggests that Uber, or others, will use advances in autonomous vehicles, renewable energy and improved transport networks to get food everywhere in the world, eliminating hunger. This might be paid for by a reduction in the military budget among leading countries. It is a nice vision, and one I would like to believe, but recent trends are towards increased military spending and reduced overseas aid, so there would need to be a major shift in politics and public opinion to see this happen.

So plenty of different styles. There must be a tendency in such situations to aim for radical ideas as a way of gathering interest and showing innovative thinking. After all nobody wants to be labelled as a boring pessimist.

3.4 Laws

Another way of predicting the future is through empirical "laws". These are observations of relationships that have held true in the past and are expected to do so in the future: hence they are not real laws, but are often treated as such by those in the industry. The best known is probably Moore's Law which has been a good predictor of the power of silicon chips over the last 50 years or more. Other laws have included Cooper's law that the capacity of global wireless communications systems tends to double every 30 months and Gilder's Law that Internet traffic doubles every 100 days.

These empirical laws can be hugely powerful and disruptive, and Moore's Law has driven much of our digital world. But such laws do not, cannot, hold true forever. Any exponential growth, such as a factor that doubles every two years, must eventually come to an end. Even if economically viable and delivering consumer benefit, in the end there are just not enough atoms in the universe.

An example of this is the size of the hard disk in the average home computer as shown below.

Learning from past predictions

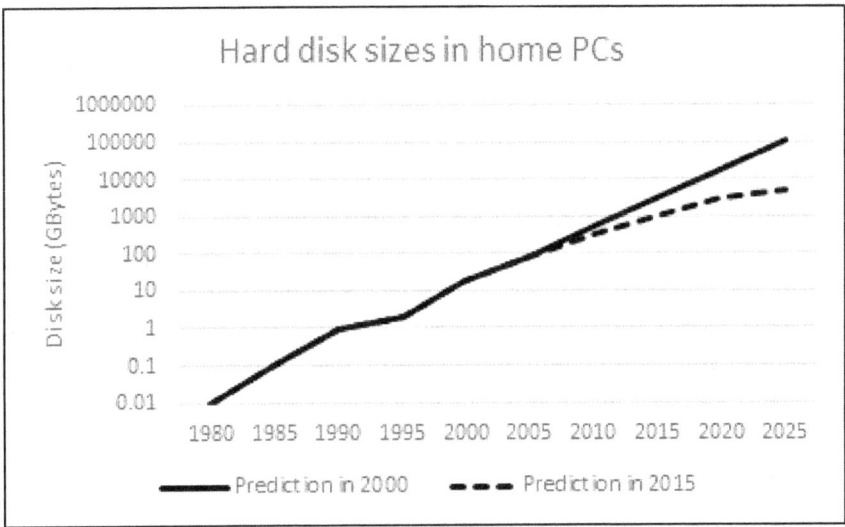

Figure 3-1 - Hard disk sizes in mid-range home PCs

Up until around 2005 a linear prediction made on a logarithmic scale was a good estimate. Hard disk size tended to grow 10-fold every nine years. But that has been slowing recently. A best fit curve in 2015 shows a plateau steadily being reached at around 10 terabytes. The average home PC owner just cannot find enough material to store to need much more, especially if they have to pay more for the privilege. It is possible, of course, that we are seeing a temporary lull, as occurred around 1990-1995, and growth will return to trend, but unlikely in that few now see hard disk size as any sort of constraint.

The same will, inevitably, occur with Moore's Law, the only question is when. The death of Moore's Law has been forecast many times, all incorrectly. But the need for larger and more powerful processors in our devices does seem to be fading. Computers are longer-lived now, and larger chips more power-hungry. With consumer demand for more processing power falling and investment costs in new chip fabrication plants ever-rising, economics may bring Moore's Law towards a plateau even if technology does not.

My feel is that, like hard drives, we are coming to the end of an era for many of these laws. Growth will continue but gradually reduce. Any predictions that rely on them continuing for 30 years or more must be suspect. Interestingly, few predictions do, AI perhaps being the exception. Even if the laws continued, in most cases diminishing returns set in, with lower marginal utility for each successive improvement.

3.5 Summary

The conclusions from this chapter might appear trite. It is easy to under-predict, especially more than ten years out, but then it is easy to swing too far the other way and over-predict. Laws are useful but need to be treated with some caution. Hot topics are often top-of-mind, but hot topics today are rarely still hot topics decades in the future. Futurologists tend to over-predict to demonstrate their credentials as innovative thinkers and because it gets more newspaper headlines. Corporate titans tend to under-predict, often because anything else might destabilise the share price of their company.

Flying cars anyone?

4 Key enablers

4.1 Introduction

Much of the change in the last 20 years has at its heart the Internet. Other key enablers have been mobile communications and to a lesser extent touch-screens. Understanding what the key enablers for the next few decades will be would make prediction a much more certain activity. In this chapter I look at those areas which seem to me most likely to enable many products and services including mobile communications (5G), the Internet of Things (IoT), virtual and augmented reality (VR/AR), artificial intelligence (AI), robotics and also areas such as the batteries which will be needed to power it all.

If an enabler is not already visible it is unlikely to change the world in anything less than 20 years (it took the Internet about that long to have any material impact). So the key enablers for the forecast periods I want to look at of 10, 20 and 30 years likely all exist. That does not mean that they are easily spotted though, and there remains a risk of something out there that has a much greater impact than currently expected.

4.2 The Internet

If the Internet has been *the* key enabler for almost all of the changes of the last 20 years, will it continue to be so for the next 20? It is certain that the Internet is not going to go away - indeed, turning it off would be nearly impossible, a fact that worries some who believe machines may take over the world. Equally, it seems unlikely that the Internet will change materially. The Internet itself is a network of interconnected computers. It constitutes the standards that allow one computer to talk to another such as TCP/IP, the hardware that enables the communications and a few key nodes such as domain name servers. It is continually upgraded to enable higher capacity and more speed, but its basic functionality remains the same. When there is a move to make a major change, such as from IPv4 to IPv6, the sheer size and distributed nature of the Internet makes this near-impossible.

When forecasters talk of change due to the Internet, it is invariably change that happens in the applications and intelligence that sit above the Internet.

It is possible that the Internet might suffer failures or become increasingly congested. Security attacks occasionally cause website failures and will likely do so in the future. But the Internet is amazingly resilient - a key design feature of the original Arpanet which became the Internet.

We can safely assume that the Internet will remain present for the next 30 years, that it will mostly, if not invariably, be upgraded as needed, but that its functionality will remain broadly unchanged. As such, it is still a key enabler, but a very familiar one.

4.3 Broadband communications and 5G

The Internet is of little value unless you can access it. Almost always we access the Internet using a wireless hop to a wireline connection. At home we use Wi-Fi to the home router which then connects to a fibre, cable or copper line. Outside the home we may use wireless to a 4G base station which then has a fibre connection to the core network of the mobile operator, which in turn connects via a gateway into the Internet. Such connections have existed for decades but their capabilities have changed substantially in terms of speed and capacity.

Back in 2000 most home connections were via dial-up modems with data rates around 30kbits/s. Mobile networks delivered a similar speed in realistic use. By 2010 home broadband was delivering in excess of 1Mbits/s and mobile was not far behind. At the time of writing in 2017 home broadband speeds were typically around 50Mbits/s and mobile data rates often similar[9]. Fibre to the home (FTTH) and 5G promise Gbits/s connectivity within a decade. Is this important?

As data rates rose, new applications were enabled. A critical point was around the 1Mbits/s mark where video transmission became possible, enabling not just streaming of video to watch but also video calling via Skype. There has not been

---

[9] Of course, this simple average hides a huge variation. Data rates in rural areas and developing countries tended to be much lower.

Key enablers

another step-change since. Faster rates enable better quality video, moving to HD and perhaps to ultra-HD (or 4k). They make downloads happen more quickly, which is nice.

Current evidence, in the form of user satisfaction surveys and take-up of faster broadband connectivity, suggests that users do not gain much from mobile speeds above around 1Mbits/s and home broadband speeds above around 10Mbits/s per member of the household. We have connectivity that is generally more than good-enough - except in those areas where coverage is lacking. On that basis, we might conclude that faster connectivity will not enable anything new in the future.

It is, of course, possible that faster connectivity will enable some new application as yet unimagined. This is an example of the "build it and they will come" thinking, which has previously been effective in building mobile communications data usage. However, it is very hard to even imagine what new applications might emerge. VR, discussed below, does not need higher data rates since it is basically a video feed, like streaming video. The IoT certainly does not. I have thought a lot and debated a lot around this topic, and while ideas are floated around autonomous cars and always-on multiple body-cams none appears convincing (a detailed discussion of the connectivity needs of autonomous cars is provided in Section 4.11).

This conclusion appears at odds with all the hype around 5G and fibre-to-the-home (FTTH) connectivity. We are promised a communications utopia where the higher speeds delivered, along with lower latency and enhanced capacity, will revolutionise our worlds. I have written about this at length in my book "The 5G Myth" but in essence it is putting the cart in front of the horse. Instead of understanding what new services we might want and then considering how to deliver them, we are providing an enhanced means of delivery and then casting around for a reason to need it. This leads to being told that we want super-high definition VR and the ability to download entire films in seconds. Developing solutions in search of problems almost always ends badly.

Hence, my conclusions that connectivity is already more than sufficient for everything we might want to do, and that improvements will not stimulate anything materially different. Innovation will come from elsewhere.

This is a bleak conclusion for much of the communications industry, although a change in strategy could mitigate the effects. At present, the mobile industry is fixated on a 5G vision that makes little sense in the hope that new concepts such as autonomous cars will need, and more critically will pay for, high speed connectivity. Section 4.11 below sets out why this will not be the case. A change of strategic direction to providing consistent connectivity everywhere, merging cellular and Wi-Fi into a seamless network and using AI to optimise and drive cost out of the network could see a return to greater profits.

The implications will be discussed further in Section 12.4.

4.4 IoT

The situation is quite different for IoT. Here, appropriate connectivity does not yet exist. IoT devices typically need to send tiny amounts of data using very little battery power and in a very cost-effective manner. Cellular networks do not meet these requirements well, and efforts to build dedicated IoT networks to date have only led to patchy coverage from competing technologies.

The problem is not technology - many appropriate standards exist (including the Weightless standard where I am CEO). It is a lack of a single agreed standard that all can coalesce around, not helped by unclear demands and business cases. These are all eminently resolvable, and will be resolved in the next five years. This is not the place for a detailed discussion of how this might occur, and nor does that matter overly for the purposes of forecasting.

Hence, my expectation is that within five years, the key enablers for the IoT will be in place. Connectivity will be widespread, chipsets will be low-cost, various issues such as security will be sufficiently resolved, data processing will be automated and business models clearer. Low data rate devices will be connected for less than around $10 including both the hardware cost and the connectivity fee. It will take another ten years or so for replacement cycles of devices until

Key enablers

IoT functionality is widespread so that by around 2035 anything that can usefully be connected will be connected.

This has the potential to be hugely important. Agriculture could become markedly more efficient. City centres could run more smoothly, with lower costs and less congestion. Healthcare could become much more proactive, rather than reactive, as more aspects of our lives and bodily functions are monitored. Life will become more convenient as things work as they are meant to, without maintenance and with on-going software updates to improve functionality. Big data analytics will generate insights that will act as a second wave of change.

The impact could be almost as big as that of the original Internet, and is explored in subsequent chapters. This is probably the most important enabler for the next 30 years.

4.5 Virtual and augmented reality

VR and AR[10] are hot topics of the moment. It would be easy to predict a future where they play a key role.

I am sceptical about VR being more than interesting entertainment from time-to-time[11]. History has shown that we do not take well to services that require us to wear glasses such as 3DTV and Google Glass. VR headsets are much worse - larger, heavier and often linked to other peripherals that you need to hold or similar. VR is also costly and often makes the user feel sick. To put up with all of that would require a compelling application and none has yet materialised. Hardcore gamers might find a role for it, and the rest of us might try it out occasionally at conferences, theme parks or similar. But expecting the average person to spend hours a day with a VR headset on seems very unlikely. When the Internet application Second Life was at its peak users were exhorted to get a

---

[10] Virtual reality is where a virtual world is presented to the user which will typically have no relation to their current location. Augmented reality is where a view of their current location, as they currently see it, is presented, but augmented with additional information such as direction arrows, or Pokémon characters.
[11] At the time of writing, Nokia agreed, shutting down its VR activities and Apple's Tim Cook had given a talk saying he felt that VR was a niche but AR was transformative.

First Life. Similarly, living in a virtual world is likely to quickly lead to a move back to the much richer real world.

I can imagine commercial applications such as allowing architects to virtually walk through buildings they have designed. There may be VR rooms in some offices in the same way that there are video-conferencing suites. But most offices do not have the space to dedicate to something like this.

Conversely, AR is much more promising. As Pokémon Go showed, even if only briefly, AR can work on existing handsets, does not require glasses, does not take users out of the real world and indeed, can help increase their interest in it. At the time of writing both Apple and Google had just launched tools to facilitate development of AR applications. AR can provide games, it can help with navigation and it can assist businesses who need to provide additional information such as schematic diagrams to their service personnel. With handsets having sufficient processing power, development suites available and apps quick and simple to develop, AR will be a useful enabler for the next few decades.

4.6   Artificial intelligence

AI fascinates and worries us in equal measures. Until recently it has been more of a dream than a reality, but major strides have been made in the last few years, predominantly in the way that machines can teach themselves using prepared data sources such as labelled images or formatted datasets. For example, the Google translate system can learn a new language simply by directing itself at sources of already translated material on the Internet. The program that beat the world champion at Go taught itself how to play by running millions of simulated games and learning which strategies worked (including developing approaches never used by humans). AI brings benefits such as advanced medical analytics but we worry that it will destroy jobs and may ultimately lead to a super-intelligence that takes over the planet.

At the moment, AI is very good at specific bounded tasks such as playing Go or learning a particular language. It is very poor at unbounded tasks such as inventing new apps or creating compelling novels.

Key enablers

AI for specific tasks is a useful enabler. It allows big datasets to be analysed and diagnostic tasks, such as medical diagnosis, to be automated. It can replace basic call-centre tasks, resolving a large percentage of enquiries itself. Current capabilities alone could lead to large-scale job losses.

AI for general tasks is far less certain. One of the ways that AI works is to train itself on existing datasets - such as pre-labelled pictures of cats when it wants to be able to recognise a cat in a picture. Such datasets exist for specific problems but are harder to define and find for general problems. Preparing appropriate "labelling" might require armies of labourer-scientists working in the modern equivalent of the mills of the industrial revolution. Where there is a widespread use of the output, such as translation, this investment in data preparation might be worthwhile. Where the output has only niche use, such as legal documents used in case law, the costs of such preparation might outweigh any benefits.

Opinion is strongly divided among experts as to whether general AI will emerge - some think it almost impossibly difficult, others simply a matter of time. Some expect that by around 2045 a singularity will occur where AI becomes so powerful it starts designing better computers itself in an upward spiral that rapidly gets out of control. Even if this process remains under our control, it might start to replace human brains in some way. This is less than 30 years away and so within the forecasting horizon for this book.

My experience in areas of such uncertainty is that they tend to take longer than experts expect and it is safer to under-forecast rather than over-forecast. For that reason, I will assume AI for specific tasks as an important enabler, but not for general tasks.

This book is generally not concerned with the impact on jobs, and in particular whether they will be replaced by other jobs, keeping employment high. However, I do touch upon this and other societal trends in Chapter 10.

One area where I expect AI to make a continued difference is the virtual assistant. This is already available in the form of Siri and Google Now and aims to provide contextual information, such as updates on travel issues on routes it anticipates you might take, calendar alerts based on how long it thinks you will need to get to the next appointment, and so on. AI will allow such assistants to learn from the behaviours of millions of people and use this to predict what you might want to do, or be interested in, at that point. This is a gradual evolution - as it has been for the last five years or so. It might result in personalised news feeds, suggested menus, recommended TV viewing, updates on favourite artists, automatically finding the best deals in areas such as insurance and even auto-responding to some emails (Google already suggests contextual-aware short responses to some emails such as "I'm not interested"). While some might find all of this intrusive, others may see it as highly convenient. For those that dislike it, there will be easy ways to disable, or scale back, its functionality.

As a way of gaining some context, in 2017 some researchers[12] took a look at the effective IQ of virtual assistants. Google featured at the top of the AI list, with an IQ of 47.2. Baidu's Duer got 37.2, Chinese search engine Sogou 32.2, Microsoft's Bing 31.9 and Apple's Siri 23.9. A six-year-old child's average IQ is 55.5, at 12 the average increases to 84.5 and at 18 it is 97. Siri still has a long way to go.

The researchers felt that the explanation for the low IQ scores might be the limited nature of most AI programmes. AI propositions are currently being built for specific purposes. AlphaGo, for instance, demonstrates a high skill level for the game of Go, but would be unable to beat a six-year-old at Hangman.

While I am generally positive about virtual assistants, some are less sure. They can be seen as "creepy", intrusive or a way of capturing an individual and their data. If there is a monopoly (eg if Google's assistant captures much of the market) there may be a backlash against what appears to be big-brother type control. Such assistants also effectively aim to make us all the same by providing help that the average person has previously needed, but most of us strongly protect our individuality. As intelligence improves the key players will need to step carefully and with sensitivity - an approach they have not generally exhibited to date.

---

[12] See https://arxiv.org/ftp/arxiv/papers/1709/1709.10242.pdf

4.7 Big data

Big data is an ill-defined term, but broadly encompasses the idea of having very large datasets from which new insights can be gleaned. The data is an output from other trends such as IoT and AI.

The promise of big data is new insights into factors that affect our lives. For example, by analysing millions of people with a medical condition and correlating with many other factors from their diet, to behaviour, to parentage, important conclusions on how to prevent the condition might be delivered. The data could be used for societal benefits such as medical or productivity, or commercial benefits such as targeted advertising.

This all sounds very plausible. However, gathering coherent datasets is not as easy as it appears and issues around privacy abound. Even when gathered, there is no promise that the insights will be delivered. But it is definitely worth a try.

Some have suggested that data is "the new oil" - a commodity that is essential to the functioning of the industrial world. However, oil is readily classified into different types and then bought and sold in liquid markets. Data is completely different. Each dataset is typically unique with its own set of fields, its own approach to gathering, and its own filtering and cleaning process. Unless the dataset is very simple, just buying a dataset is generally insufficient - you also need to understand how it was gathered and processed. This means that datasets tend to be associated with companies where the employees involved in the data gathering process reside. It is typically necessary to buy the company, rather than just buying the data, in order to gain access. Even then, end-user agreements made before data gathering could start may limit the usage of the data. All told, for larger companies, it is probably easier just to gather their own data, unless it is particularly difficult to assemble.

As a result, those most likely to benefit from big data are the companies that either already have large data sets, or can assemble them easily. Start-ups can gain some insight but will typically then be acquired. Big data is not like oil, but perhaps more like owning an oilfield - those that do can benefit inordinately and can exercise control on the downstream market.

4.8 Robotics

Robots have been part of the future for almost as long as there have been predictions of the future. They have invariably disappointed. Robotics is a tough challenge. Interest in robotics is currently high, especially as a way to supplement aging populations in countries like Japan. Here, it is hoped that robots could become carers for the elderly living at home, taking on household tasks and also acting as a companion[13].

I have already discussed, and shown scepticism towards robots that undertake tasks around the home and similar. Robots that can provide companionship are entirely possible. Indeed, smart speakers such as Amazon Echo are effectively already starting to deliver this sort of functionality, and AI will help here. Whether this will be an enabler for much more, or just a sticking plaster for loneliness is unclear. Robotics also raises many ethical and societal questions such as whether we are content to let robots take human jobs, whether it is humane to use robots as carers, whether we should let robots bring up our children and much more.

4.9 Batteries and power

More and more of our devices are battery powered. Keeping phones, computers, wearables and even toothbrushes charged is a growing challenge of everyday life. In some cases, people reject new devices in their lives if it means yet another thing to charge. In other cases, applications are constrained by battery capability - for example AR applications running on handsets are very power hungry, needing to keep screen, camera and processor all running at full power, often for extended periods. Bad battery design can lead to devices catching fire with resulting safety and reputational risk.

Battery technology is critical for electric vehicles. It could revolutionise renewable energy by enabling storage of solar power during the day for delivery at night.

---

[13] While logical, this raises many ethical questions around caring for others which, I expect, will engender much debate over the coming decades, slowing the advent or robots in such roles, and potentially even blocking them.

More generally, many aim not to increase their personal environmental footprint and so would prefer that new services and products did not require ever-more energy. Always-on devices are often viewed with suspicion even if their standby drain is low.

Batteries then are more of a constraint than an enabler.

Batteries have evolved steadily over the years. But breakthroughs are rare and becoming increasingly unlikely. We can expect costs to fall, especially for the sort of batteries needed for electric vehicles, as mass production leads to efficiencies. Broadly, this makes little difference for my forecasts. Indeed, whether cars are electric or petrol is of little relevance for the digital future - we will still have cars (and they won't fly, especially if they ned to carry batteries with them).

I expect batteries to remain a constraint into the foreseeable future. This may limit our inclination to adopt additional devices or power-hungry applications.

4.10 Autonomous vehicles

After years of trials, in 2017 autonomous vehicles appeared nearly ready for mainstream use. It is clear that cars will take over ever-more driving functionality. Already, features like adaptive cruise control and automatic emergency braking are widespread, meaning the car effectively controls its speed. Lane maintenance is simple on motorways and major roads. The key question then is not if autonomous vehicles will happen, but by when, and by evolution or revolution.

Google's view is that it will be a revolution, with vehicles moving directly to full control. However, there are many good reasons to think evolution is more likely. All sorts of unexpected behaviours will happen, for example other drivers pulling out in front of autonomous cars safe in the knowledge they will stop. Today's traffic often relies on slightly aggressive tactics, pulling out a little into the road to encourage oncoming vehicles to give way. European city centres are vastly more complex affairs than the relatively open roads of California which are used to test most autonomous vehicles. Then there are myriad issues around liability in accidents, questions around the extent to which drivers need to be able to take over control and much more. These will take decades to fully play out in a safety-conscious and highly regulated environment.

An interesting example of how autonomous cars might struggle with city centres is a pedestrian crossing. Cars are supposed to give way when people want to cross. Human drivers are very good at gauging whether someone near the crossing is likely to step out. They can assess, for example, whether a pair of people are chatting and stationary, or both about to step out, from the way they are looking and their general demeanour. Autonomous cars will just see a couple of people and will likely stop as a precaution. In a busy city there may be a steady stream of pedestrians walking past the crossing and the autonomous car might be stationary indefinitely, waiting for them to clear. This sort of example suggests that city centres may need some re-design to enable autonomous cars, replacing crossings with traffic-light controlled systems, reformulating complex junctions and perhaps segregating cyclists and drivers. This would be a work of many decades.

I expect to see automatic lane maintenance shortly, autonomous driving on motorways in a few years, extending to out-of-town roads in 5-10 years and then gradually to more complex environments. But it might well take the full 30 years of this forecast until cars are truly autonomous everywhere and are widespread.

4.11 Autonomous vehicles and 5G

As discussed above, autonomous vehicles are slowly progressing while the mobile community is pushing ahead rapidly with 5G. Some believe that the two are complimentary. In this section, I explore this in more detail as a way of demonstrating why the form of connectivity promised for 5G might be of limited use and why autonomous cars will be slow to emerge.

A car might need many different forms of connectivity. A telemetry connection would enable uploading of engine and vehicle information and the download of software updates. This would typically only need to happen infrequently, such as daily, and could be achieved using cellular and even satellite for the downlink. Passengers might require connectivity for entertainment and work purposes and this is delivered today either direct to the devices using 4G or via a Wi-Fi

Key enablers

repeater in the car, often working from a 4G connection on the roof. Finally, autonomous vehicles might need map updates, congestion information and perhaps control information. Telemetry and passenger communications are already achieved with existing connectivity, it is the autonomous operation that is considered here.

Two extreme views on autonomous operation could be imagined. At the one extreme, autonomy really does mean autonomy. The car is on its own with no connectivity, only the pre-loaded maps, and it navigates itself around, using on-board sensors to make decisions about accelerating, braking and direction. Let's call this "truly autonomous". At the other extreme there is no autonomy and the car is under the complete control of a network directing its every move. Let's call this "network control". Current autonomous cars are closest to truly autonomous whereas the future envisioned by the 5G community is network control.

The truly autonomous car is clearly possible as is currently being demonstrated in trials over millions of miles of driving. Is the network control car any better? Full network control is clearly not viable. This would require connectivity across every centimetre of every road, otherwise the car would need to come to a halt and await human intervention. Such connectivity does not exist now and is unlikely to do so in a 5G era, where the higher radio frequencies deployed will result in much less coverage. Indeed, low-latency 5G connectivity might only be available in city centres. Hence, all cars need to be truly autonomous.

This leads to a refinement of the question - is a car with occasional network control better than one without network control? "Better" in this context means bringing benefits that outweigh the costs to the owner of the vehicle. Proponents of the network control car mainly argue that it is safer, with a secondary argument that it might allow greater road capacity through platooning and other congestion management techniques. Opponents note that the connectivity needed will be extremely expensive and the central computer system needed to control a nation's cars is massively complex, liable to have bugs and need funding and regulatory approval. Low-latency connectivity then will be only available in dense areas and expensive. Let us now look at safety in this context.

Since autonomous vehicles will spend most of their time being truly autonomous (because low-latency connectivity will not be available on much of their route) then they will clearly need to be very safe. Google and others have already demonstrated levels of safety well beyond those of human drivers and this will likely only get better both as algorithms improve and as the percentage of autonomous cars goes up. When collisions do occur, they tend to be at low speed. How might network connectivity help? Imagine a case where a child runs out into the road after a ball. A centralised network will not know about this so the first car will need to detect the child using its sensors and take evasive action. It will likely warn cars behind with vehicle-to-vehicle (V2V) signalling so they will know about it. But a car coming around the corner, too far back to see the V2V signal, might be surprised on encountering stationary vehicles. In principle a network could warn it about this. In practice, a car that ensured its braking distance was always less than its forward visibility would not need such a warning. Of course, there are many more scenarios to consider, but it is hard to see network connectivity as providing a material improvement in safety, and certainly not one users will pay hundreds of dollars a year for.

So why does the 5G community believe that autonomous vehicles are an important demand driver for 5G connectivity? As I have already noted, 5G, to some extent is a solution in search of a problem. Those designing 5G have postulated that, just like every previous generation, faster speed is essential. Current 4G systems can deliver 100Mbits/s in good network conditions, 5G aims to raise that beyond 1Gbits/s. Higher speeds also bring lower latency (because each bit is transmitted more quickly). The 5G community asked itself what applications might want such high-speed low-latency connectivity. The existing customer base of consumers does not appear to need it; few consumers notice any improvement in connectivity when speeds increase above 1Mbits/s - a factor of 100 less than currently achievable. A new "customer" base is needed that requires this connectivity, has sufficient volume to make the hugely expensive network deployment worthwhile and has the economics to pay the monthly charges needed. The mobile community decided that autonomous vehicles came closest to meeting these requirements and propounded the myth that autonomy required network control with centimetre-level precision. They have extoled this view with such conviction and so frequently that many have come to believe it.

Key enablers

Solutions in search of problems often end badly and this looks to be no exception.

4.12 Quantum computing and security

Quantum computing is an idea that it is very hard to get your head around. Quantum computers can computer all different possible answers to a question simultaneously, making the resolution of some problems massively quicker. In quantum entanglement changing the state of one particle causes the state of an "entangled" one anywhere in the world to instantly change. Quantum computers are steadily developing in laboratories but are incredibly difficult to make.

Perversely, quantum computing could both compromise security and make security finally completely robust against all attacks. As a result, it might be both a disabler and an enabler.

Security is a potential threat to any digital future. Hacking, denial of service, phishing and many other attacks continue. On the Internet we are mostly protected against them via a mix of anti-virus software, rapid program updates and individuals learning about scams. In the IoT security breaches are concerning and might well cause many to think twice about buying that connected kettle. Firms frequently suffer data breaches and lose data with the result that passwords and identities can be compromised.

My view, in general, is that Internet security is somewhat like concerns over road accidents or even health concerns over mobile phones. We know that there is a risk of injury or death while driving but accept this risk because the benefits of mobility are so great. Likewise, some have nagging concerns of the health risks of mobile phones (despite overwhelming evidence to the contrary), but still use their phone because they could not imagine life without it. We continue to use the Internet despite knowing there are security issues because we view the risk that they will affect us as minor and the benefits of the Internet huge. Security concerns will continue, but the Internet community will react swiftly and resolve each one. I do not expect such a major breach that the Internet shuts down, or that we decide to stop using it.

Quantum computing might change all of that. In principle, a quantum computer would be extremely good at performing the calculations needed to break the encryption we use on the Internet - so-called public key cryptography. In the wrong hands a quantum computer might threaten to crack any Internet transmission leading us to stop using it. The solution is to use quantum cryptography. Using quantum entanglement, messages could be protected in such a way that they would be utterly safe against all attacks. More simply, using a quantum "fingerprint"[14] of a chip can provide excellent security. If quantum develops there may be a race between decryption of existing messages and encryption with new quantum techniques.

Quantum computing might also help with AI and other complex problems by bringing a step-improvement in processing power. Whether this will be usable, and advantageous remains to be seen,

It is very unclear how quickly quantum computers might develop, whether they will ever be viable, and whether they will be able to perform useful calculations. In 2015, experts at Microsoft predicted that quantum computers would be available by 2025. My best guess is that quantum will develop more slowly and will not materially change any forecast here, with early computers being of limited use. But this is another area of substantial uncertainty.

4.13 Blockchain

At the time of writing there was much hype around blockchain. From its beginnings as a means of enabling currency without banks, it has grown to be seen as the answer to almost any problem and a key enabling concept. I am sceptical.

Blockchain effectively allows the disintermediation of a trusted entity. In the case of currency, the banks can, in principle, be dispensed with. There may be other areas, perhaps medical records or similar, where the ability to share

---

[14] Each chip has very slight differences as a result of an imperfect manufacturing process. These differences result in slight variations in voltage and current levels within the chip. By characterising each chip as it comes off the production line and storing the characteristics, someone receiving a message can check that it came from a particular chip, and therefore a particular device. This idea is already in use.

information in a trusted way, either allows for disintermediation or opens new opportunities.

Set against this is the complexity of blockchain and underlying issues such as the energy requirements needed to maintain them (the "mining" process). Disintermediation can save money - as it has done with the Internet. However, removing entities which exist because they are trusted is much more problematic that those there simply to match suppliers and customers. Also, such disintermediation may lead to unintended consequences, such as facilitating the laundering of money by criminals. I doubt blockchain will have a major impact on our digital future. If it does have an impact, then it is too early to predict with any confidence what it might be.

4.14 Summary

The key new enablers identified in this chapter were IoT, AR and AI. Useful advances may occur in big data and robotics. Autonomous vehicles were seen as somewhat important. Previous enablers such as the Internet and broadband connectivity have reached a point where they are not enabling anything new and so are unlikely to drive further innovation.

In subsequent chapters I move on to provide predictions for a range of different environments. It is worth taking stock before I do this as to what the first four chapters have shown.

From the world of magic and science fiction I concluded that key areas where there was an unmet need that seemed plausible were:

- Even better language translation.
- Robots with more intelligence.
- Micro-robots - but it is not clear what they would be used for.
- Replicator-3D printer.
- AI.

Looking at past predictions suggested that while 50 years ago there may have been a tendency to under-predict, that has swung the opposite way in recent times and over-prediction is more likely. The most current issues also tend to have too great an influence.

There are a few areas of uncertainty, primarily around how fast AI will evolve and whether it will be able to leap from specific tasks to general tasks. Perhaps most surprising was just how few truly uncertain areas there were.

5 The home

5.1 Introduction

Ideas for the smart home have been around for many decades. The picture below is from the 1960s and shows a smart vacuum cleaner.

Figure 5-1 - The smart home of the 1960s

Almost every prediction of the future has assumed homes that automatically do things for its occupants, from cleaning to security to comfort. And yet almost none of this has come about. The smart fridge has become a standing joke for all that is wrong in futuristic products.

The typical home in developed countries does contain much in the way of innovative technology. Most have complex routers, Wi-Fi nodes, streaming TV services, multi-room sound systems and more. We will, it seems, pay for entertainment, sometimes thousands of dollars for a larger TV screen. All of this is quite well illustrated with the chart below, based on data from Deloitte[15].

---

[15] See http://www.telegraph.co.uk/technology/2016/08/27/internet-of-things-struggles-as-

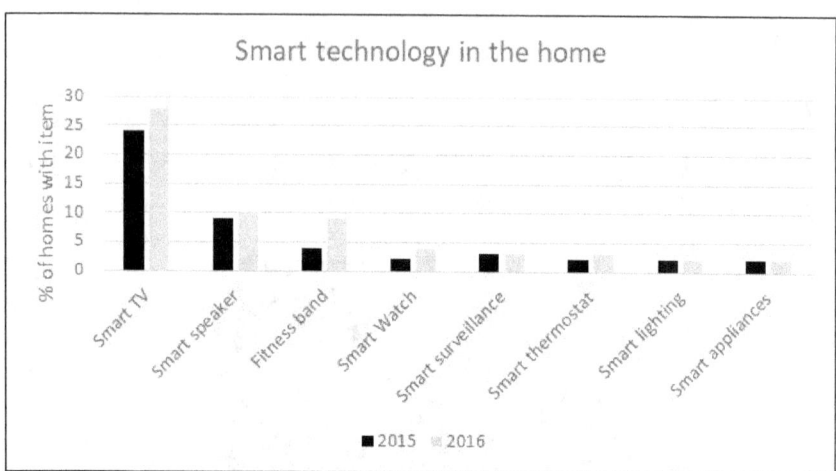

Figure 5-2 - Penetration of smart devices in the home (in the UK)

The figure shows relatively high and growing penetration of entertainment-related devices, but low, and mostly static penetration of "smart home" devices.

Home ownership may change over this period. Globally there is a slight trend towards renting, and this might be accelerated by the sharing economy and perhaps variants of Airbnb that cater for long-term occupation. If this were to happen, it would lessen the desire to change the fabric of the home by either the owner or the occupants.

How then to predict an area that has generally disappointed in actually delivering?

5.2 Home automation

There are many ideas for better homes - almost all available as products today. They include:

use-of-smart-home-gadgets-flatli/

The home

- *Robotic vacuum cleaners.* Available for a decade or more, but with virtually no market penetration because of the issues discussed earlier such as an inability to climb stairs.
- *Smart lighting systems.* Automatic lighting is widely deployed in office buildings where workers cannot be relied upon to turn it off, but rarely in homes. Systems can change the colour of the light and enable iPad control but are expensive, proprietary and seen as a gimmick by most.
- *Connected security systems.* Many homes have burglar alarms and some are connected to third-party security centres. Increasingly mobile phone alerts are available. However, this just allows you to watch while your home is burgled! Smoke detectors are sometimes connected mostly for maintenance reasons to ensure they are still working. Electronic locks are fitted in commercial buildings (eg activated with a pass-card) but keys work remarkably well for most residences.
- *Smart HVAC[16].* Systems such as NEST have become widely deployed, enabling remote control of heating and air conditioning. This can make sense in energy savings by leaving the system mostly off and manually activating when needed. Linking the heating control to a user's diary is still some way off[17].
- *Smart appliances.* Fridges, washing machines, dishwashers, even coffee machines, are gaining Wi-Fi connectivity, but benefits are unclear. Smart kettles, for example, seem pointless, since you need to go to the kettle both to fill it with water and then to pour out the boiling water.
- *Smart toilets.* Discussed further below.
- *Smart metering systems that measure gas and electric use.* These are becoming widespread, mostly due to government initiatives, but suffer the "two weeks to the drawer" problem where users find the information interesting for a week or two then get bored and put the

---

[16] Heating, ventilation and air-conditioning - an acronym often used in the US. Others simply refer to the home heating system or similar.

[17] For someone living at home, with a simple diary (eg always leave for work at the same time), then the prediction task is relatively easy, as long as the person keeps their diary up to date. But imagine a house with four occupants, some of who work shifts, others travel frequently, some come back from University in the holidays and so on. Not all maintain a fully comprehensive calendar and some use Google, others Apple. This becomes a challenging task which likely involves learning about the typical patterns of each person.

display unit in the drawer. Few change usage patterns significantly as a result of smart metering.
- *Plant watering.* Soil moisture sensors are available which can alert home users that their plants need watering. However, a lack of an alert might be due to battery failure as much as not needing watering. Most can spot a wilting plant and take appropriate action without the need for a device.

Showcase homes of the future have had all of these, but as noted, very few real homes have many of them. There are many reasons, but broadly all are expensive and complex and the benefits are not seen as worthwhile by most.

For some concepts the home fabric itself needs alteration. Anything requiring wiring is expensive, often needing floorboards lifted and holes drilled in walls. For this reason, current solutions almost invariably use wireless connectivity rather than wired connection. For robotic vacuum cleaners, strips may need to be laid under the carpet to mark "no go" zones. Smart HVAC typically requires a trained electrician to install.

New homes ought to be better placed in that necessary wiring can be installed during construction - but it is far from clear what this might be. Homes still have standard telephone sockets and TV coax cables installed when these are rarely used. There was a trend to fit Cat-5 (Ethernet) computer wiring to hi-spec homes, but few would now connect to this, using Wi-Fi instead, or Ethernet-over-powerline solutions as needed. It seems whatever is deployed will become redundant within a few years. This has led to homes being built ever more "naked", with the inhabitants choosing to clothe the home with their preferred digital products and using wireless to link them all together.

In some cases, the product feels too proprietary. Smart lighting systems typically need a home controller and light bulbs all from the same company. Having installed an expensive solution there is a risk that the company goes bankrupt, or decides to discontinue that product - it has happened many times in the past. I am trying to replace a fault smoke detector only to discover that the model is now obsolete and I need to replace the entire home system of connected detectors - with a new unit that needs a different mounting bracket and connector.

Insecurity is an increasing concern. There has been recent publicity about hacking into smart kettles and then being able to gain the Wi-Fi password and eventually control of the home network. While security breaches are rare, few want to put themselves at risk unless the benefits are clear.

Device replacement cycles in the home are long. Fridges last for 15 years or more, washing machines, vacuum cleaners and so on for 10. Few would upgrade before end-of-life just to get smartness, so changes often take a decade, during which time the world has moved on in any case.

Keeping everything working is also a headache. Systems do break, and often the fault diagnosis can be difficult. They are expensive to replace. There is much to be said for the simple light switch.

The biggest problem is the lack of sufficient benefit. Do you really need to be able to see the inside of your fridge while away from home? Is a message from the washing machine that it has finished much use when you will go and empty it whenever convenient anyway? Will avoiding the minute wait for the coffee machine to turn on make your day happier? And while technology might improve and costs might fall, there is no obvious way to enhance the benefit. Without that, home automation is destined to remain something we see in "homes of the future".

A good way to end is with the smart toilet that can assess human waste and provide medical diagnosis and dietary advice. This was predicted earlier in one of our set of predictions provided by visionaries in Section 3.3. Technically, it is reasonably straightforward. But would anyone go to the trouble of taking out existing toilets, installing new ones that likely need power supplied and other modifications to the home fabric and maintaining them (they will probably need to be topped up with certain chemicals and perhaps cleaned occasionally)? Only if the benefit is large. Most of us have managed quite well with the occasional trip to the doctor when something seems wrong and an annual health check-up. Only the paranoid, the already ill or those with particular propensity towards certain health problems would seem to need this. Having said all of that, the Japanese already tend towards electronic toilets to a greater extent than many other countries, and might be drawn to this sort of functionality first.

5.3 Home entertainment

The opposite is true of home entertainment - progress has often been faster than predicted. It has also been very difficult to predict. A decade ago we expected most video consumption on large screens whereas it now occurs on tablets and mobile devices. We thought content would be stored in a central home server rather than the cloud. We anticipated a single family-wide home system whereas in practice some elements are shared, other individual. We expected a dedicated sound system rather than streaming from mobile phones to connected speakers. Gaming seemed destined to need more space and equipment, but if anything has tended in the opposite direction.

If few can see the benefit of home automation, it seems most are sold on even the smallest benefit from entertainment. Upgrading to a slightly larger TV, perhaps with a curved surface, is seen as important. The latest home TV box is a necessity. Higher definition video is always a good thing (although higher definition audio appears less important). Perhaps it is because the entertainment system takes a central role in the home and can act not only as a source of pleasure in its own right, but like an expensive car as a form of social prestige. So where next for home entertainment?

Visions of the entertainment system fitted into the home fabric are fading. Once, people predicted TV screens integrated into the wall, or even the wall being one large TV screen. Art hung on the wall would actually be TV screens in picture mode. TVs would be integrated into bathroom mirrors, speakers into every room. As with home automation, this is just too difficult. Running the wiring is expensive. Mounting the TV on the wall requires DIY skills. Far easier just to put it on a stand against the wall. With massive TVs available, there is little need for a whole wall to become a screen - and if it did it would probably break at some point. Speakers can be discretely located and do not need to be built into the ceiling. We can expect that our home entertainment system will sit within the home, as opposed to being part of the home. Not only does this make sense for home owners, it is essential for those renting homes and those with relatively short occupancy such as students.

Entertainment is also tending to the individual. While families still do value a room where they can gather to watch a programme together, watching is clearly tending towards tablets and similar. This means a house-centric entertainment system rarely makes sense, instead content needs to be stored remotely (eg in the cloud) and pulled down by the individual. So home hard-disk content storage devices are likely to become less useful and will go the same way as DVD players.

As already discussed, visions of home holographic displays are fanciful. They are not practical and in any case, do not fit the model of individual consumption. A home VR unit is possible but like the Wii or X-Box would not become a dedicated part of the home, rather it would be a box connected to a broadband line and the headset and other peripherals stored in a drawer in the room.

All the home really needs is good broadband, good Wi-Fi (or home Ethernet), a few screens and speakers and a simple way for phones and tablets to connect to local resources. Less is more.

5.4 Family living

In the Weasley's home the Whereabouts Clock tracked the family. Some visions of home living have allowed for various enhanced forms of family interaction such as integrated calendars, family chat systems and smart screens in the kitchen that post family-related content.

Most of this is available already on handsets and tablets. Adding an additional tablet just for the kitchen is a luxury – and one that requires charging, maintaining and replacing as it becomes redundant. Easier to bring a personal tablet into the kitchen and call up these various features as needed, while still being able to monitor personal social media.

At one point around five years ago, Microsoft demonstrated a wonderful coffee table where the central part was both a screen and an image recognition system. It could call up content, enable interactive games and more. But it never made it out of the concept stage. It is hard to see many paying likely $1,000 or more for something that just makes it easier or more fun to do things that are already possible. And getting power to a coffee table which tends to be in the centre of a room is difficult.

5.5   IoT and AI

In Chapter 4 I noted that key enablers included IoT and AI. Both would initially appear to have strong application to the home. Devices like Amazon Echo might be early markers of what is possible, both enabling conversational interaction with people and connecting up devices around the home. At the time of writing there was vibrant competition among major companies to lead on the home smart speaker and its related eco-system. Many would extrapolate this to a fully smart home.

But the need for such devices remains unclear[18]. Many are used just as a convenient way to call up music. Despite promotional offers and relatively low prices (compared to, eg, a large screen), sales are small to date.

Part of the problem is that there is little for these devices to control. As noted earlier, not much in the home is smart, is wireless enabled, nor is likely to be. Despite valiant attempts to deliver smart toothbrushes, smart forks and smart bathroom scales, there has been little take-up.

The Amazon Echo is so cheap that homes may buy one just as a way to play music in a room. Once adopted, it may be that people find a role for it, or that social media adapts to work well on that format. Really simple integration of connected devices might reduce some of the barriers to bringing smart technology into the home. While it is a beguiling vision to tell Alexa to turn off the lights, and it is eminently possible, it does not seem to be addressing a real need.

This is an area of some doubt. Leadership from Amazon or others could result in low-cost and readily connected devices that met the requirements of the home.

---

[18] A recent description of the latest Amazon Echo was that on saying "hello Alexa" in the morning it could turn the lights on in the kitchen and turn on the kettle. However, these are not tasks that most find troubling!

5.6 Predictions

On the one hand, there are many areas where home automation could make our lives somewhat better. On the other, predictions of smart homes made for over 60 years have consistently over-predicted the pace of change. Which way to lean?

I have concluded, after having over-predicted this area myself in the past, and having looked closely at why that was, that despite all predictions to the contrary, homes are unlikely to change much. The benefits just do not add up to enough to persuade home owners to make the investment. I would rather it was not so – I love gadgets as much as any engineer.

My predictions are:

1. Automated home HVAC systems will become ubiquitous in 10 years during which time they will also gain the intelligence to adapt automatically to the diaries of the home occupants.
2. Smart speakers such as Amazon's Echo, will be widely deployed within five years, but mostly used as a kitchen radio.
3. Around 20 years from now AI will have evolved to deliver some useful in-home features that are not currently clear to us today.
4. Most innovation will occur at the device and the individual level rather than at the home appliance and household level.

6 At work

6.1 Introduction

The corporate HQs of large multinationals often feel as if the future has arrived already. Automated systems log you in. Conference rooms have tablets outside the door with the room booking status. Lights come on and off automatically. Doors require pass-cards or other forms of identification. Rooms routinely have large screens or projector units, conference phones and touchpad controls. Bathrooms have automated toilet flush, soap and water provision. Accessing the Wi-Fi is appropriately complex requiring host agreement. Plenty of stainless steel and glass completes the picture. The end result is somewhere between Minority Report and 1984.

The dynamics here are different from the home. Automation often makes commercial sense since any activity requiring humans involves cost. With no natural home-owner, systems to ensure policies are followed and intrusions minimised make sense. Buildings are somewhat frequently refurbished, allowing for new cabling, and the structure of modern buildings often makes under-floor cabling simple to install and to provision to the right place in the room. Even if the cost-savings are insufficient, the brand value of appearing to be leading-edge can be valuable for some corporates.

This appears much more fertile ground than the home. In this chapter I look firstly at office buildings, then at remote working, then at other working environments, before examining the influence of our key enablers.

6.2 The office

Offices can be expected to introduce anything that makes commercial sense. Broadly, this means using technology to remove the need for staff – whether they be receptionist, cleaners, IT specialists or others. Salaries tend to rise steadily whereas technology tends to get cheaper, so more technology can be expected to appear in the future than in the home. I look at a number of different "topics" as a way of structuring the discussion.

At work

Biometrics are important for companies. Identifying individuals allows them to open doors, access IT systems and so on. At present, this tends to be pass-cards that are held against sensors, but in the future more advanced biometrics are likely, using facial recognition or similar. Cameras, widely distributed, might be used for a range of monitoring purposes.

IoT could be widespread. Plant soil monitors do make sense in a company where people might not notice wilting plants, or might not take ownership of watering them even if they do. Waste bin monitors can help with emptying on time. Coffee machines might be monitored for maintenance. All of this will make the office work better rather than adding new functionality.

Buildings already have a small amount of intelligence – for example they turn lights on and off (although many will have experienced lights that turn off while they are still there, but have not moved from their desk for a while). AI could be helpful in keeping the lights on more accurately, adjusting the heating and so on.

Robotic devices have got more chance in offices. Many tend to be open plan, with large floors and do not have stairs (or could have a separate robot per floor). This makes robotic vacuum cleaners relatively simple. They can come out at night and do their work. Robotic waste bin emptying is plausible although desk cleaning might be a step too far. Some maintenance might be possible – for example IoT could be used to inform of malfunctioning lighting and robotic devices could fix the problem, but in practice the labour saving here is probably too small to be commercially sensible (and maintaining the robot might be more effort than replacing the bulbs).

All of this might be mostly unnoticeable to the workers in the building. It is designed to make the environment work better, rather than bring any new features.

Office IT has evolved over time. Many now bring their own devices (an approach often known by the acronym "bring your own device" or BYOD) and expect them to connect via Wi-Fi. Office phones are disappearing as workers use their mobiles, use Skype or similar. Most now just need a clean desk and a place to plug devices in.

The exception to this is connecting to printers (becoming less frequent) and to large screens for presentation purposes. Interestingly, predictions of the future made 20 years ago forecast that laptops would connect automatically and wirelessly to projection screens. Sadly, it has not happened. Getting connected to the projector is still a complex task, firstly involving finding an HDMI adaptor from whatever version is used in the plug in the laptop to whatever version is provided on the end of the cable. Then it requires getting the projector to recognise the laptop and the laptop to operate in dual-screen mode. Sometimes all this happens automatically, but it is rare. So why is this? It appears to be predominantly a lack of standards. With many different laptop manufacturers and many different screen manufacturers, getting every device to work with every other one requires well-defined standards not just at the physical level but also at the software level. The incentive to resolve the problem is not really there – few choose a laptop on the basis of whether it will connect easily to other screens. You can imagine that if Apple made the laptop and the screen it would all work beautifully: but they don't (yet).

This might seem a trivial discussion. But it does hint at possible issues with more complex IT integration, not just in the office but elsewhere. Wherever products from different manufacturers need to come together there is a risk they will not work. In some cases, such as with Bluetooth, the standards community has resolved these issues over time. But in others, where the connectivity is less central to the device function, it might not be worth the effort of fully sorting the problem. Alternatively, it might hint towards a future where we pick our favourite manufacturer and buy all our products from them (a future Apple would approve of).

At the moment, some large corporations in their premium conference rooms have complex room IT systems, with bespoke control units and clever integration of the screen, conference phone and more. I have not come across anyone who likes these. They take time to understand and often are not intuitive. Every system seems to be different. They must be expensive and yet they just annoy people. So either they will get more intuitive or they will stop being deployed. The experience of connecting to a projector suggests the latter.

6.3 Remote working

Workers are not always in the office. Past predictions suggested that they might never be, working from home most of the time with excellent video conferencing facilities. This book is not primarily about sociological factors, but perhaps it is just worth noting that the reason this has not happened is not "digital" – the connectivity needed is now widespread. It is social – people like interacting face-to-face with others. They like getting out of the home, even if they do not like the commute. It is commercially sensible – people work better when they interact directly, teams are stronger, innovation improves as we chat over a coffee. Remote working will remain a minority occupation.

Remote working can use broadband connectivity, Skype (or other video conferencing systems such as Zoom), document sharing in the cloud and so much more. It is hard to see what more could help.

6.4 Non-office working

Not everyone works in an office. Here I consider some of the larger non-office categories of work. Of course, there are many, many more types of work than listed here, but this set should provide a flavour of the ways that digital might change working arrangements in the coming decades.

6.4.1 Agriculture

Agriculture adopts new technology where there is a business case. If a technology can improve crop yield or reduce the manual intervention needed then it is often implemented. In the last decade many of the advances have been in farm machinery, which can drive itself, deliver fertilizer or other products with precision and make measurements. Drone and satellite imagery has also proven useful at spotting areas of fields that needed intervention or different treatment. The advantage of approaches like drones is that one drone can provide feedback for the whole farm whereas monitors in the soil or crop are needed at regular intervals. Livestock monitoring is starting to occur. Cows and other large animals are valuable, and monitoring them can detect ill health, the right timing for insemination and an understanding of where they go on the farm. Automated feeding systems and milking systems can save on labour and are slowly being introduced.

A key change could be the widespread introduction on IoT monitoring. This might be soil monitors distributed every few meters, sensors on plants, overhead sensors in greenhouses, etc. Such sensors could enable high-precision agriculture where water and nutrients are delivered almost per plant. Systems might first find deployment in high-value crops, such as grape vines for up-market wines, and then be steadily introduced in other areas. It may be that these systems have the biggest impact in developing countries where aspects such as irrigation are especially critical when the water supply is scarce. The success of IoT depends in part on whether low-cost, simple IoT solutions can be developed and proven to enhance yield sufficiently.

The net effect would be a steady improvement in agricultural efficiency and perhaps a steady reduction in cost, making food cheaper. The number of people employed in farming might fall, although the impact on harvesting might be small.

6.4.2 Vehicle maintenance

Vehicle maintenance might seem a strange area to select, but in many developed countries, the motor vehicle sector (which includes sales as well as maintenance) is the largest single source of employment.

Car maintenance has come a long way in the last few decades from greasy garages where diagnostics was based on experience, to shiny automated dealerships where the first step is to plug the car into the diagnostic system and subsequent work is prescribed by a computer. Maintenance is more likely to simply involve swapping out parts rather than repairing them.

This trend could continue. Cars could get ever-better at monitoring every key element and raising alerts as various items wear. Equally, cars are now astonishingly reliable and it is unclear whether there is much to gain. Monitoring an already reliable system might run a higher risk of the monitoring system failing and triggering false alerts than any material gains in reliability being achieved.

In principle, robotics could help in standard servicing tasks such as changing the oil. The task is harder than on an assembly line where the car is placed perfectly, all parts are new and the number of model variations very small. Also, a dealership will see far fewer cars than an assembly line, so the amortisation of the cost of the robot would be across a much smaller number of cars. This suggests little, if any, application of robots at this level.

Much more likely is that electric cars will dramatically reduce maintenance. Electric cars have far fewer moving parts and electric motors need no servicing. Widespread use of electric vehicles would cut the number of people needed in maintenance. But hybrid electric/petrol vehicles are more complicated, with two drive-chains. Hence, we might see a world where for the next decade or so the maintenance requirements for cars grow, before falling in subsequent decades. Some Governments have declared that they would like to see all cars being electric by around 2040 – although the history of such central "decrees" is that they are quietly dropped if it proves commercially difficult. If a move to electric does happen then 20 years from now most cars will be electric and by 30 years all will.

Car sharing might reduce the number of cars on the road. It is already happening in large cities and would be much facilitated by autonomous cars that could move themselves to where they are needed. This trend also looks likely to cut vehicle maintenance needs in the 20-30 year timeframe.

6.4.3   Retail

For a while the writing on the wall appeared to be clear for retailing. Internet commerce grew inexorably, resulting in the decimation of high streets and steady bankruptcy of venerable shops and even newer entrants – Toys R Us had just announced their demise at the time of writing. Internet retail is more convenient, lower cost and offers wider choice.

But high streets have survived through a process of reinvention. Some people enjoy shopping still ("retail therapy") and high streets, or shopping malls, offer a venue for a day out, mixing browsing the shops with a moment of luxury in a coffee shop and a chance to spend time with friends and family. For clothing and other goods some prefer to touch the product and try it on first, but developments in virtual fitting where users can see a visualisation of themselves wearing the clothes may change much of this. Flagship stores can be useful for branding purposes. While we clearly have not yet reached the end-state it will clearly involve some stores remaining while most utilitarian retail is completed over the Internet.

The key effect is on jobs, with retail shrinking from one of the largest employment sectors. Some of these jobs have shifted to warehouses, web site development and delivery, but these are also at risk. Warehouses are becoming automated – this is an environment where robots can be used albeit picking some products that are awkward shapes is problematic[19]. Delivery may become more automated although there are many difficulties here. Driverless cars have already been discussed and mostly discounted, and in any case, someone needs to get the parcel from the vehicle to the front door of the house, knock on the door and then get confirmation of delivery. Drones are talked about but could only be used for the lightest of parcels and face the same delivery issue, as well as struggling to find somewhere to land in dense urban areas.

6.4.4 Construction

Another large sector is construction. Here digital has had less impact. Construction workers still lay bricks by hand, much the way they have for centuries. Of course, digital helps with the building design, and new materials are being introduced in many areas. Prefabrication and on-site assembly is tried from time-to-time but generally seems not to be preferred. Some suggest houses could be 3D printed, but that would seem rather fanciful (and need a very big printer).

The lack of impact from digital to date suggests a lack of impact in future. Robotics generally fails on muddy construction sites where each building is different although it can be used in pre-fabrication factories. IoT can help with building maintenance where monitoring devices are embedded into the fabric of the building, but this just makes the construction task harder. AI has no obvious role.

---

[19] Over time, packaging will likely standardise on certain form-factors that robots can handle.

6.4.5 Hospitality

This includes hotels, restaurants and similar. It is an area where digital has had a minor impact. Some hotels have dispensed with receptionists, with room keys provided by a check-in machine and billing performed automatically (simplified by removing all optional extras such as mini-bars from the room). Equally, others see an imposing reception as part of the experience that travellers are paying for. But most aspects of hotels and restaurants remain very hard to either automate or replace by robotics. Many staff are relatively low-paid decreasing the commercial incentive to automate their tasks away.

Some predictions of the future (and I erred here too) have travellers arriving at hotel rooms where their phones automatically set air conditioning to their preferred levels and other similar tasks. While it is now normal to connect to the hotel's Wi-Fi (albeit after having found out what the password is, or having accepted terms and conditions), linking into the room's controls has not happened. The benefits are minimal – it is not hard to press a button on the room thermostat system – and the complexities and risks high. In particular, any system would need to be secure to prevent pranksters hacking into a hotel and making all the rooms freezing cold. This is a similar problem to the office presentation projector discussed earlier and similarly unlikely to be resolved.

Finally, for many, hospitality is something to be experienced, not a parcel to be delivered. The enjoyment includes the interaction with the receptionist or the waiter. Too much digital might make the experience worse rather than better.

6.4.6 The digital factory

Manufacturing plays less of a role in employment than it used to, especially in the developed world. Partly this is a move of jobs to lower-cost countries and partly an increase in automation. We would not now expect to see a car manufacturing plant without robots festooned across the production line.

While the change has been gradual, consultants such as Accenture now postulate we are on the brink of a revolution which they term "Industry X.0"[20]. In such a

---

[20] This term grew from the concept of naming the Internet in various stages. The first Internet was one where we accessed information. The second - Internet 2.0 was where we

world most things in the factory are automated. Designers, perhaps in different countries, use computer design tools to come up with a new product which is sent to the factory's computer system and then assembled automatically. They note that the key issue in such a world is not the availability of robots, but the skills and culture of the workforce. Instead of teams of relatively low-skilled assembly workers, higher skilled digital designers and robot maintenance staff are needed. This, in turn, needs a different management structure, different employee incentives, different HR system and so on. It is almost a completely different company and changing from one to the other, especially for well-established manufacturing players with long heritage, could be extremely difficult. Equally, the barriers to entry for new players are very high: they need to build robotic factories, establish a brand (eg for consumer white goods), build the expertise in making high-quality product and deliver at scale. This makes the transformation slower than it might be as culture takes years, even decades, to change, without the forcing function of strong competitive pressure.

Looking at the longer term, the trend is very clear. Robotics will be increasingly introduced on production lines, initially in the form of static units with a moving production line, but eventually with moving robots able to be more flexible. Factories will be designed to facilitate robots, not humans, and products designed to be simpler for robots to build. Only in the very specialised, such as high-end Swiss watches, will humans continue to manufacture. This will be a very gradual process, as it has been up to now. Each factory will make a decision on timing based on the relative costs of human labour compared to robots. For the developed world the change may be in the next decade. For the developing world in might be 20-30 years away.

6.5 The changing nature of work

This might all be irrelevant if we no longer have jobs. One of the key questions that is often posed about AI is whether it will make some, or all of us, redundant. In such a vision we will lead lives of leisure (or depressed unemployment depending on your world view). Towards the end of this book I will make predictions for the role of digital in our lives and then extrapolate from these to

---

shared our own information using social media and similar. Some have suggested we are now at the third, or fourth, but it is unclear what the major distinctions are. Hence, the move away from a specific number to "X.0".

broader social implications such as employment, albeit that will not be the focus of this book.

But to avoid suspense, I do not believe we will all stop work, nor that the nature of businesses, the need for head offices and so on, will change materially. So looking at the role of digital in the workplace is worthwhile.

6.6 Predictions

The office potentially will see much more digital because it makes commercial sense. However, we might not notice since the aim is not to add new functionality but to reduce maintenance costs. Specifically:

- IoT will progressive be deployed in lights, plants, waste bins, coffee machines and more, starting within 5 years and extending to 15 years from now.
- Biometrics will be increasingly used to open doors and enable IT systems, starting almost immediately and extending to 10 years from now.
- Robotics will make some impact, with robotic vacuum cleaners and possibly other solutions to tasks that need to be performed daily. This will start 10 years hence and extend to 30 years into the future. (The reason I suggest it will take 10 years is that there has been no real usage so far, despite the availability of robots.)
- AI will be used to optimise aspects such as lighting and heating, but this is a relatively minor application. This will not be where AI shows its true capabilities.
- IT in the office will nearly all go away as BYOD predominates. This is already well in train and will complete within 5 years. Even conference room IT systems will broadly disappear with simple TVs/projection units left for slideshows.
- Of the major employment sectors agriculture will see some steady productivity improvements continuing current trends, vehicle maintenance will grow over the next 10 years but then decline rapidly afterwards, retail will continue to decline at approximately its current rate, construction and hospitality will be largely unaffected, and

manufacturing will see ever-growing penetration of robots on the production floor.

7 Travelling

7.1 Introduction

We typically spend around an hour a day commuting or travelling for other purposes. Travelling is often fraught. Trains delayed, traffic jams, queuing at airports. Almost all predictions of the future magic away the pain points, with concepts such as automatic notification of delays, and diversions around congested areas. Some also assume that the transport itself will be better – faster, more refined, better connectivity, etc.

Much of this has happened. Most cars now have satnavs with traffic avoidance (although, of course, if all take the same route to avoid congestion it just shifts the congestion elsewhere…). Live train and plane departure information is available on-line. Trains and planes increasingly have Wi-Fi connectivity (and cars are even getting there). Intelligent assistants (Siri and Google Now) can work out the most likely mode of travel to get to the next destination and then check on delays. Ticket purchase is becoming automated by touching a credit card against a reader.

Equally, it is important to recall that when presented with a choice, most of us opt for the lowest cost airline and accept long queues at the gate and less facilities on the plane. Just because something is possible does not mean we will adopt it – broadly it needs to be nearly zero-cost.

Some push back. We like holidays on canal boats travelling at walking pace. Sometimes a longer journey is seen, rightly, as a chance to reflect, away from the constant digital stimulus. For many, the advent of Wi-Fi on planes is a bad thing.

7.2 Planes

If I were writing this in the 1970s I might conclude that planes were set to get faster. Concorde had just been introduced into service and was clearly the future. But 40 years later, planes are no faster than they were before Concorde. If I had been writing 10 years ago I might have concluded that the future was bigger – the A380 was on the horizon. But at the time of writing orders had dried up. The

Boeing 737 was still the workhorse of the industry. Despite many experiments, economics has forced us back to the same sort of planes as the 1970s. It is hard to see this changing.

In-flight entertainment *has* changed. From viewing a fixed program on a screen at the front of the cabin, longer-haul flights now offer seat-back displays with on-demand films, or the ability to connect a BYOD to the screen to view pre-stored content. Like home entertainment, it is hard to see how this might get better beyond larger screens – but the size of the seat-back clearly limits this.

There is room for improvement at the airport. While check-in has largely been dispensed with, and boarding passes are now downloaded to phones[21], security is still unpleasant. However, resolving that seems unlikely. Airports are, rightly, hugely risk averse, and threats evolve. The perfect system might just scan us and our bags as we walked along an open corridor. Such a system is far from realisation. Bag tracking could be improved, with the ability to see exactly where a bag was - but this would be more reassurance than an improved service.

Overall then, little change is likely.

7.3 Trains

Like airlines, stepping back 20 years and we might have imagined the future of train travel was magnetic levitation and bullet trains, travelling at 300mph across the country. Some trains have improved in speed a little, but the constraints of the train tracks and growing congestion have largely prevented speed increases.

A big improvement is the increasing availability of Wi-Fi on trains. Even if we are not travelling faster, at least we can be connected. At present, the Wi-Fi is constrained by the backhaul from the train carriage to the network, often via 4G cellular. This can be expected to improve over time, using microwave dedicated

---

[21] This is another long-standing prediction, but one fulfilled in an unexpected way. Forecasters (including me) expected the phone to use Bluetooth or near-field communications (NFC) to communicate with the gate boarding check. In practice, we display the boarding card on the screen and use a visual reader. Much simpler in that it avoids connections needing to be established albeit more intrusive for the traveller who needs to get the boarding pass onto the screen and present the phone in the right manner.

transmitters. The technology is available; the key question is who should pay for the service.

Ticketing is increasingly contactless "touch and go", although finding the best prices often involves pre-booking via the Internet.

As with planes, it is hard to see what further improvements are practical. The speed of the trains, their size and form factor and the cost of service is unlikely to change. Some operational efficiencies are possible with IoT devices monitoring track and rolling stock, reducing maintenance costs and failures which ought to lead to somewhat reduced ticket costs. Driverless trains are also eminently viable but the savings, in the overall scheme of operating a railway, are not huge.

Perhaps the best we can hope for is that our trains will become ever-more punctual (expect in Switzerland where they are all on time already), a little cheaper, and that we will have broadband connectivity when aboard.

7.4 Automobiles

I have already covered autonomous cars in Section 4.10 but there is more to say about factors such as entertainment.

Prior predictions have often focussed on the car navigating itself, providing hands-free communications capabilities and perhaps sending telematics back to the manufacturer for maintenance purposes. Much of this is now available, although often with little to do with the car. Guidance is provided from a smartphone sat near the dashboard. Entertainment is direct to the smartphone via cellular or using the cell-phone as a Wi-Fi hotspot to repeat a signal to other devices in the car. Hands-free operation of the phone can work well direct from the cellphone, although the car's audio system is often a little better. Bluetooth links allow the car to talk to the devices, and it could use them to send back telemetry, although this rarely happens, perhaps as much for security reasons as anything else. In a way, this is an echo of the home, where the individual is more important than the family – in the car the individual is more important than the passengers collectively and hence the preference to use individual devices rather than those of the car.

Cars of the future will be amazing pieces of technology. From headlamps that can stay on high beam but "notch out" particular parts of the beam to avoid blinding on-coming cars, to monitoring the state of the driver, to predicting and avoiding collisions, the level of innovation is breath-taking. All to be replaced eventually by autonomous vehicles which do not need any of it.

The biggest problem for those using cars is typically congestion. The digital world might go some way to resolving this. Smarter cars might have fewer collisions and break down less, reducing accident-induced traffic jams. Platooning might increase road capacity by allowing vehicles to drive closer together. Centralised journey planning could allow drivers on routes that would otherwise become congested to depart at a different time or take a different route. However, this will be offset by growing numbers of vehicles on the roads in many countries and increased commuting distances if journey times fall.

Finally, car ownership may decline. Sharing already allow for car pools, where a car is used only when needed. Technology, and especially autonomous vehicles that can drive themselves to where they are needed, would speed this (and blur the line between shared cars and taxis).

7.5   Radical

Transport always attracts radical, science-fiction like ideas. At the time of writing Elon Musk had just proposed rocket-transport between major cities, enabling trans-Atlantic travel in 30 minutes, and suggested it could happen in less than seven years. Elon has an astonishing track record, achieving the near-impossible with Tesla and SpaceX, so his perhaps his proposals should not be immediately dismissed, but that is exactly what I am going to do - this just seems ludicrous to me.

I have already talked about flying cars and magnetic-levitation trains. These are not new ideas, yet no progress has been made for decades. No new technologies have arrived which are game changers. Concorde showed that too few would pay for speed to make it economic. Any bets on radical transformation of transportation would be very long bets indeed.

7.6 Predictions

Predictions of the future often find ways to solve our current travel inconveniences and frustrations. Daily news stories discuss autonomous cars, rocket-transport and sometimes jet-packs or hover-boards. Surely, the future must involve better transport?

But think back. Transport has not changed materially for decades, other than in better connectivity when travelling and sometimes simpler ticketing. When we have a choice, we opt for cheaper rather than better (think low-cost airlines). Digital does not change the underlying nature of mechanical devices.

My prediction is very little change. There will be better on-board connectivity – we are already well on the way to achieving this and it will broadly complete in the next 5-10 years. We will have simplified ticketing and payment – again this is already in place on buses and underground trains in many countries so extending it to other areas is a relatively minor task.

There may be some improvement in congestion through digital scheduling of journeys and notification of alternative routes. But this may be offset by increasing population. There might also be some productivity improvements through IoT reducing maintenance requirements, autonomous trains, and so on, but these are relatively minor savings and will likely make little difference to the end-user price.

8 Leisure

8.1 Introduction

We spend as much, or more, of our time at leisure as we do at work. If the predictions of AI and robotics reducing employment are correct, we may be spending even more at leisure. This chapter looks at how digital will change our leisure activities.

8.2 Cycling as an example

The types of leisure are many and varied. Examining each one would not be possible. Here I start by picking on one that I know well- cycling - and then extrapolating from this to the more general.

About 30 years ago there was no technology involved in cycling, other than that used in the manufacturing process. Gradually bike computers then made an appearance, measuring speed and distance and increasingly other parameters. By 2010 bike sat-nav devices were widely available and route-planning on the Internet with download to the device was possible.

In the last five years technology has exploded. On the Internet this has been led by sites such as Strava – a kind of Facebook for runners and cyclists. Strava allows you to upload rides, set targets and track progress against them, gain insight into levels of fitness, compare yourself to others, chat to other cycling friends and so much more. On the bike the more advanced sat-navs can link to radar detection systems, to automated lights and to heads-up displays in glasses. Electronic gear shifting systems can be programmed to work in just the way that suits you. My bike, which is far from the latest, has three wireless sensors measuring speed, cadence (pedalling rate) and power. I wear a wireless heart rate monitor. The information all gets sent wirelessly to the Garmin head unit which combines it with information from GPS and Glonass satellite systems then links via Bluetooth to my mobile phone, which in turn uses 4G to maintain network connectivity.

In training, virtual environments can be conjured up when on static trainers, allowing you to race in the virtual world against friends or against virtual training partners. Cycle fit sessions on specialised units allow every aspect of your posture and fit to the bicycle to be analysed with millimetric precision.

Of these, by far the most popular has been Strava. Partly this is because it is free, with premium features available for a relatively small subscription fee. Partly it is because it meets a basic desire to show off about a great ride, and get "kudos" from friends. Partly because route planning is useful and setting and meeting targets can be genuinely motivational. Many (including me) can get quite upset if a ride does not upload, and there is a saying that "if it's not on Strava you didn't ride it" which reflects the human need for progressing against targets and for gamification to be a form of social glue.

The adoption of bike technology has been mixed. Most cyclists will have a sat-nav, although some find their phone mounted on the handlebars adequate. The heads-up display has not found favour, nor the radar detection systems. Electronic gearing is quite well received, but some feel it takes away the tactile feel of the connection to the bike.

Of course, riding a bike still involves pedalling, often getting wet and sometimes falling off. Bikes still look the same and despite all the claims to the contrary, my experience is that the underlying bike has not changed materially in the last ten years (the biggest change of significance was the introduction of carbon fibre frames about 20 years ago). But the cycling experience has changed notably with Strava and sat-navs. Strava continues to add features. AI is also likely to help, generating optimal routes, analysing the performance of millions of cyclists and learning which behaviours give the best results and in other ways we cannot imagine. Cycling is better for having had a digital makeover and likely to get better still in the next decade.

8.3 Enhanced leisure activities

Cycling may provide a guide for the more general impact of digital on existing types of leisure. Digital may not materially improve the underlying equipment such as the cricket bat or tennis ball. It will provide analytics, social sharing, training advice and virtual ways to undertake the leisure when it is not possible to do it in the real world. With apps being relatively easy to write, specialised websites and apps will emerge for each different class of leisure, likely using the same underlying platforms or code blocks. This makes leisure more enjoyable and also extends its scope. Now I can play at cycling not only when I am out on the bike, but also at home, comparing rides on the Internet, using social media to discuss with friends and building training plans for the next ride.

Leisure appears to be tracking somewhat behind work and entertainment in its adoption of digital. We had Facebook a decade ago, but it is only in the last three years or so that "Facebook for cycling" has emerged. It feels like mainstream digital has stabilised, with few new concepts either on the phones or the apps, whereas new concepts bubble over for leisure activities. This may be because leisure comprises hundreds, perhaps thousands, of specialised markets, which are less lucrative than the mainstream because of their size, and require specialised knowledge to address. Only as the tools to build apps and websites have improved to the extent that it is relatively easy to build, for example, customised social media applications, has it become practical to address leisure areas.

Hence the world of leisure seems set for significant change over the coming decade as the Internet adds value to most activities in the form of insight, analytics, social media and organisation of activities. This can also be extended to include retail which, as mentioned earlier, is becoming more a form of leisure than a task to be completed. While this will not change the fundamentals of the activity it will make it more pleasurable and able to soak up more time.

8.4 New forms of leisure

Digital has generated whole new genres of entertainment, primarily in the form of digital gaming. From simulations to "shoot them up" to puzzles, new forms of entertainment fill our hours. These games are now a massive industry and are clearly going to remain a core part of our leisure for the foreseeable future.

New types of entertainment are possible. AR/VR aims to deliver a mix of improved traditional games (eg making shooter-games more realistic) and also new experiences such as virtual tours of real venues like museums. As Pokémon Go showed, AR can mix together gaming and more traditional pursuits (such as going for a walk). AI could provide conversational-level voice interfaces which might make new concepts possible or existing concepts much simpler. The smart speaker is an early example of this, using voice for web interaction.

8.5 Predictions

In summary, I predict that:

- Each form of leisure will get its own specialised digital enhancement affecting the equipment used through to the supporting websites and apps. This will burgeon over the next decade but then plateau.
- Digital entertainment such as games will continue their steady growth towards ever-greater realism and complexity, adding voice control and other AI-related features.
- New forms of digital entertainment, enabled by AR/VR and similar are likely, but it is unclear whether these will be interesting additions or will take over existing markets.

9 Public services

9.1 Introduction

In developed countries spending on public services is often 40%-50% of all national expenditure. Key public services include healthcare, education, law and order, emergency services and defence.

Digital has been much slower to make an impact in the public arena. Initially, this might seem surprising, as I will discuss, the logic to move to on-line education is compelling and the impact that digital could have on the delivery of healthcare vast. However, the dynamics of change are often radically different in the public sector for the following reasons:

- *Risk aversion*. A public-sector career can be seriously impacted by a high-profile failure but only marginally improved by a great success. Civil servants quickly learn it is better to guard against failure than to push for innovation.
- *Lack of profit motive*. Public services are centrally funded. They need to balance their costs and income but are not there to make a profit. If they over-spend Government will typically have to provide them with more money, albeit perhaps after replacing the departmental head. If there is private sector competition this is often seen as helpful, reducing expenditure.
- *Bureaucracy*. Public bodies are typically large entities. They have complex procurement rules and processes designed to protect against individuals making inappropriate choices. They have annual budgets that might mean any new concept has to await the next budget cycle before being funded.
- *Lack of innovative individuals*. Generalisations always mask exceptions, but those drawn to public service are often individuals who prefer a predictable and secure employment. This means innovation typically needs to come from outside.
- *Politics*. In some cases, even where a digital service makes perfect sense, the politicians would find it difficult to introduce. For example, a

digital product reduced the need for nurses might be spun by journalists as a callous Government sacking hard-working nursing staff. Better to leave things as they are.

The question to consider in this chapter is whether this state of relative digital backwardness will persist, perhaps indefinitely, or whether after decades of falling behind, public services are now poised for major change. I will consider this by firstly examining two of the largest sectors which have the greatest impact on most citizens (compared to, eg defence or prisons).

9.2 Healthcare

In 2016 the IBM AI program called Watson demonstrated that it could perform diagnosis of illnesses at least as well as doctors, often better[22]. At the same time, doctors were often awaiting letters from consultants sent through the post. It all sounds crazy.

There is so much that digital could do for healthcare. Monitoring of individuals could happen in the home, tracking vital signs, the proper taking of medicine and general well-being. Initial diagnosis could be mostly automated. Hospital admittance, often taking an hour or more to gather information from patients, could happen on-line beforehand. Medical information could be immediately available on-line to professionals throughout the system. AI could assess outcomes from hospitals and professionals and advise on best-practice, as well as raising the alert for poor performance. It has even been shown that robots can perform health consultations not only effectively, but be preferred by some patients as an easier way to discuss embarrassing issues.

All of this would both result in a higher standard of healthcare and a much lower cost. Patients could be kept out of hospital longer and expensive professionals only deployed where absolutely needed. Will it ever happen?

The areas where digital has had the most impact are those where individuals can take control. People buy healthcare products for the home that help them live

---

[22] See, for example, https://futurism.com/ibms-watson-ai-recommends-same-treatment-as-doctors-in-99-of-cancer-cases/

longer alone, such as smart scales, fall alarms and more. However, these are restricted so, for example, anything that monitored blood sugar and prescribed insulin would not be available without full approval from the "healthcare system". Where there is a profit motive, such as with health insurance in the US, in some cases health insurers will look for digital solutions that can cut their costs. At the other extreme, incredible technology is used in some leading-edge operating theatres under the control of skilled surgeons.

Efforts to introduce new systems have typically failed. In the UK an ambitious project to deliver a single IT system and database of patient records across the entire national healthcare service (NHS) was finally abandoned after years of delays and cost overruns and under political pressure for cost savings (and political embarrassment at spending "patient money" on expensive IT consultancies). Where systems are introduced they are often second-rate due to the procurement process and fostered on staff who are typically not particularly IT literate with insufficient training and support. This makes matters worse rather than better and reinforces the view that digital and healthcare are poor bedfellows.

Finally, the human touch is important. Sometimes the role of a GP is to act as a confident, to sanction rest so the body can heal itself, and to provide sympathy. A health service which treats patients as if they were cars coming to a garage for repair might well result in a population that became less well due more to psychological than physical effects.

This has all the makings of irresistible force meeting immovable object. The pressure to implement digital will only grow, not only in terms of meeting budgets and needs, but also as patients see the increasing mismatch between what they know is available and what their doctor can use. But the mechanisms to introduce change will remain the same, with all the same barriers. If it were the private sector a new entrant would simply come along and revolutionise the system. But in healthcare, regulations prevent that happening – as does the overall funding structure: VCs typically do not have the patience to wait for all the various trials and setbacks needed. If revolution cannot happen then it must be evolution – albeit slowly. The world is a big place and some countries will be more amendable than others to digital. Success stories might spur others to follow. Patients having better diagnostic tools than GPs might put public pressure onto politicians for change. The economics of an aging population might leave us with little choice than to innovate at a greater speed than the system would normally like.

In summary, digital will change healthcare by a slow, evolutionary process. It will take 20-30 years and may forever be playing catch-up with the wider world.

9.3 Education

Over the last five years there has been much interest in massive open online courses (MOOCs). MOOCs allow students to study at home, streaming lectures over the Internet, accessing course material and submitting assignments. Some Universities offer degrees based on remote study.

Of course, none of this is particularly new. In the UK the Open University (OU) accepted its first students in 1971 –nearly 50 years ago – and has flourished since then using and pioneering this formula and keenly adopting new technologies as they arise. But it educates less than 1% of the university students in the UK.

Tuition fees are rising around the world while the through-life earnings increment from having a degree is falling. In places like the US it is unclear whether going to University makes financial sense any more. Surely the time for MOOCs has come?

This argument presupposes that the main reason for going to university is to get a degree that will increase earnings potential. But for many it is much more than that – a life experience, a way to meet new friends and try new activities, a chance to mature before entering the job market, and a way to "learn how to learn" in a world where during their lifetime many will end up in jobs not yet invented. It is an insurance against a world where MOOCs may remain being seen as second best by employers who themselves went to university. Perhaps that is the problem for MOOCs and similar, their only benefit is lower cost rather than a better education and so they are seen as second-best. Shifting that impression will be almost impossibly difficult.

For some second-best may be all that is possible. If university study is too expensive, or impractical because of, say, the need to care for elderly parents or young children, then online is the only approach. Cost factors may be particularly relevant in developing countries where the local education may be sub-standard but the costs of travelling to a developed country to study prohibitive.

Like healthcare, slow evolution is likely. Lecturers increasingly make their lectures available on-line so students may stay away more, studying them, or indeed the best-in-class versions from around the world, at home. Term times might shorten, reducing costs. Working while studying might become more practical, strengthening the links between business and education. Lecturers might like this, gaining more time for their research activities.

Of course, education starts much earlier than university. MOOCs for five-year olds do not sound like a good idea; at that age school is as much about learning behaviours and social skills as it is learning content. Younger children need supervision and parents are rarely around to provide it. If nothing else, school is a cost-effective form of childcare. There has been evolution – for example some homework is performed on-line and marked automatically, both saving teacher time and providing immediate feedback to the student.

Mainstream education seems unlikely to change much but the flexibility for those who cannot engage in the normal manner, will be valuable.

9.4 Predictions

We have seen a very different world in the public sector – one where in many cases digital would make a material difference but has failed to be introduced because of all the various barriers to adoption. This means most of the public sector is significantly behind the private sector in its use of digital. Much change is possible, even inevitable, but only slowly. We can expect the use of digital to continue to gradually transform public services, perhaps taking 20 years or more (after all, they have not changed much in the last 20). Many of the transformations will be the same as we have already seen in the private sector such as on-line access and the steady application of AI.

Public services

We are unlikely to see pioneering innovation in the public sector, although in some areas, such as robotic care assistants, new concepts might emerge. This is unfortunate as it is an opportunity missed.

10 Structure of society

10.1 An interim summary of our predictions

In this chapter I am going to look at how digital has affected society to date, and then discuss both how it might affect society in the future, and how society might affect the introduction of digital. In particular, I want to consider whether a backlash against some of the manifestations of digital, such as zero-hours contracts, will slow further changes. Before I start, I summarise the predictions I have made without consideration of the impact society might have on them.

I have suggested that overarching themes and key enablers *might* be AI, IoT, robots and smarter digital assistants. I have noted that there is greater likelihood of over-prediction than under-prediction and that we are all too willing to believe that the world is an ever-faster changing place.

I believe homes will not change much. I anticipate widespread adoption of automated HVAC systems within the next decade. I also expect widespread adoption of smart speakers but that they will be little used beyond being smart radios. Innovation will mostly occur at the device and the individual level, although AI might provide some useful capabilities in the longer term.

I expect the office to change much more, mostly because it makes commercial sense. However, we might not notice since the aim is not to add new functionality but to reduce maintenance costs. The office could gain IoT functionality in areas where it removes the need for people, including biometrics to enable access, all within the next decade. It might see robotic functionality such as vacuum cleaners in the 10-30-year time horizon. It will lose integrated IT systems as BYOD predominates and even complex conference room systems will be removed.

Outside of the office there will be IoT deployments leading to productivity improvements in agriculture and robotics deployments enhancing manufacturing. Construction and hospitality will be mostly unaffected, while car maintenance will grow in the next 10 years and decline after that.

Structure of society

Regarding transport my prediction is very little change. There will be better on-board connectivity – we are already well on the way to achieving this and it will broadly complete in the next 5-10 years. We will have simplified ticketing and payment – again this is already in place on buses and underground trains so extending it to other areas is a relatively minor task.

Leisure will evolve, with IoT making leisure devices better, and the Internet improving social interaction, skills and planning of leisure activities - essentially specialised versions of Facebook for each leisure activity. A few new forms of leisure based around AR and VR games might emerge but will only form a small percentage of total leisure activities.

We have seen a very different world in the public sector – one where in many cases digital would make a material difference but has failed to be introduced because of all the various barriers to adoption. This means most of the public sector is significantly behind the private sector in its use of digital. Much change is possible, even inevitable, but only slowly. We can expect the use of digital to continue to gradually transform public services, perhaps taking 20 years or more in areas such as on-line access and the steady application of AI.

So broadly, the largest changes will be at work, in various work environments and in our leisure activity. Other areas might see little change.

10.2 Society today

We live in a world that has realised most of the dreams of science fiction writers and avoided most of their nightmares. If our world was described to those living 100 years ago – less than a lifetime for some – it would seem a magical utopia. Food delivered to the door at the press of a few buttons on a magic tablet. An infinite variety of entertainment. A world where few die from accidents and working conditions are mostly pleasant.

Despite all this, we are not generally any happier than in previous decades, and perhaps less happy. Mental health issues abound with frightening statistics that around a quarter of all teenagers now have some form of mental illness. Obesity affects a large share of the population and shows no sign of solution. Protest votes against the status quo have led to anti-establishment leaders such as

Donald Trump and votes to leave the European Union in the form of Brexit. Economic growth is elusive and the future seems ever more uncertain as not only jobs for life, but jobs at all, disappear or morph into zero-hours contracts. Inequality is growing is many countries, and digital is seen as a key reason for this, not just in the vast wealth that the successful entrepreneurs accumulate but in separating the "haves" - those with digital skills - from the "have nots" - those with less relevant skillsets.

Blaming digital for all of this seems far too simplistic. However, our best guess at the cause of the mental health issues is the impact of social media both in setting impossible standards and in making children more reclusive. Obesity is partly due to sedentary lifestyles which digital mostly encourages. Zero-hours contracts have been enabled by the Internet and connectivity. Digital has removed jobs in areas such as retail and threatens more in the future. Digital enables "fake news" and other mechanisms that many believe facilitated the election of Trump and the rise of extremist political movements. Many see inequality as a symptom, if not a cause, of much that is wrong in society, and digital has certainly aided inequality by making it simpler for knowledge workers to accumulate wealth but harder for those without digital skills.

It would be quite reasonable for society to slow digital change while all of this was sorted out. But will it?

10.3 A digital backlash?

Digital currently feels a little like indoor plumbing. Some might want to return to simpler times but decide against it when they realise that past times involved outside toilets and a lack of central heating. Equally, surveys show that over a fifth of the population would rather give up sex than their Wi-Fi connection. For many the mobile phone is the most important thing in their lives. Perhaps that is a symptom of a problem with digital - that it is a form of additive drug.

At the time of writing it did feel that society was starting to push back in some areas. Zero-hours contracts were being outlawed in some countries; others were demanding better employment contracts for such workers. Global digital companies such as Google were being required to pay tax in local countries. At the time of writing Uber had just been told it could no longer operate in London,

Structure of society

although a resolution to the problem is likely. Privacy and data protection laws are due a major upgrade with the introduction of the European General Data Protection Regulation (GDPR) in mid-2018.

Understandably, AI is the area that evokes most interest and concern. Issues raised include:

- Will AI make me redundant?
- Will I have to interact with "bots" when I'd rather speak to real people?
- Will AI evolve to a level where we can no longer control computers?
- Can AI programs be (inadvertently) discriminatory, perhaps on a gender basis?

Some recent studies have suggested that AI will destroy jobs. For example, PWC came up with the following[23]:

Employment shares and the estimated proportion of jobs at potential high risk of automation by early 2030s for all UK industry sectors

Industry	Employment share of total jobs (%)	Job automation (% at potential high risk)
Wholesale and retail trade	14.8%	44.0%
Manufacturing	7.6%	46.4%
Administrative and support services	8.4%	37.4%
Transportation and storage	4.9%	56.4%
Professional, scientific and technical	8.8%	25.6%
Human health and social work	12.4%	17.0%
Accommodation and food services	6.7%	25.5%
Construction	6.4%	23.7%
Public administration and	4.3%	32.1%

---

[23] See http://pwc.blogs.com/press_room/2017/03/up-to-30-of-existing-uk-jobs-could-be-impacted-by-automation-by-early-2030s-but-this-should-be-offse.html

Industry	Employment share of total jobs (%)	Job automation (% at potential high risk)
defence		
Information and communication	4.1%	27.3%
Financial and insurance	3.2%	32.2%
Education	8.7%	8.5%
Arts and entertainment	2.9%	22.3%
Other services	2.7%	18.6%
Real estate	1.7%	28.2%
Water, sewage and waste management	0.6%	62.6%
Agriculture, forestry and fishing	1.1%	18.7%
Electricity and gas supply	0.4%	31.8%
Mining and quarrying	0.2%	23.1%
Domestic personnel and self-subsistence	0.3%	8.1%
Total for all sectors	100%	30%

Table 10-1 - Employment areas at risk

This suggests 30% of all jobs are at risk, with transport, retail and manufacturing high up the list. Perhaps this is no surprise – retail and manufacturing jobs have been falling over many decades in any case. But increasing unemployment by 30% would radically change the structure of our society.

Of course, a 30% increase in unemployment is unlikely. Firstly, the estimates from PWC may be an overstatement. Secondly, previous experience of job losses in one sector is that employment grows in different areas. After all, something like three quarters of the population was employed in farming at one point, only around 1% is now, but we do not have 74% unemployment.

My view is that both are true. The estimates are likely to be an overstatement because they likely assume autonomous vehicles, advanced robotics and other factors I have discussed earlier which, for various reasons, will not transpire in a way that replaces people in many cases. Secondly, we currently have relatively

Structure of society

low unemployment despite the job losses that have already been occurring in many of these fields, suggesting that jobs are being created elsewhere. Equally, I am not complacent. While no sociologist, I wonder whether we are reaching a point where our consumption is sufficient. There are only so many meals out, Netflix movies and large TVs that a person can use in their lives, and with few new things emerging to drive another wave of consumption, it may be that unemployment does grow. Some would argue this is a good thing – that we are reaching the point that Keynes suggested in his paper on "economic possibilities for our grandchildren" where we only need work 15 hours a week in order to afford everything that we want.

Whether any of this is true or not is better addressed by others more qualified to speak on this complex topic than I am. I am more interested in whether there will be a Luddite-style backlash against AI. I suspect not, most accept that pushing back the tide of digital progress is nearly impossible. Taxi drivers might demonstrate against Uber but few have much doubt as to where it will all end.

What about the other concerns raised? Some are bothered about being forced to speak to a bot when they would prefer a human. This may be akin to the off-shoring of call centres. For a while companies pursued it as a way to save costs, but eventually realised that they were losing customers as a result. Most call centres have now come back on-shore. Broadly, if enough people do not want to interact with a bot then competition will ensure that they do not have to. It would only be in areas where there is no choice, such as interacting with the national health service, that this would not be true, but even here citizens have voice through their politicians.

AI is likely to have unintended consequences, such as discrimination against certain groups or types of people. Like all unintended consequences, all that can be done is to spot and correct them as quickly as possible.

Overall, it seems likely that while there are many concerns, and some that dislike the digital revolution, the key action is to temper its impact through tools like employment legislation rather than to attempt to stop it, or even redirect it.

10.4 The impact of society on our predictions

In summary, society is unlikely to have much impact on the predictions that have been made here. Many of the future enablers such as IoT and AI are seen as good things, and supported by Governments with grants and research funding. Many countries desire to be leaders in a digital future and push strongly in this direction. By 2037 almost everyone will be "born digital" - having had digital surround them from birth, and will probably be less inclined to push back against it in search of a past they never experienced. The digital divide will shrink and digital will be seen like plumbing - something that has been there for as long as anyone can remember.

Even if a country did decide that it was uncertain about digital progress, it would be unable to do much about it. Most digital services are inherently global and hard to block at a national level (although the Chinese try hard). Some specific companies, such as Uber, could be barred at a country level but unless most of the population is in agreement that this is a good idea, Governments will be unwilling to go down that route.

Closest to a current impediment is the much restrictive regulation around privacy, embodied in the European GDPR legislation. This will make it harder to store data, limiting big data analytics and the ability of AI to use data for training purposes.

There will likely be ethical concerns in the future around AI, the role of robots in healthcare and more. This is nothing new for society – there are currently much more difficult ethical questions around DNA manipulation. But it is something new for digital.

Overall, I see no need to change my predictions in the light of societal issues. This is not to say that societal issues are unimportant, far from it, just that changing the digital future is not the best way to address them.

11 Predictions

11.1 Introduction

In my book "The future of wireless communications", written in 2000, I said of 2020 that we would have delivered almost all the services and technologies that we could envision that we need. After that would come a period of slower change, partly because there was no need for new concepts, and partly because society would need time to digest all the new tools at its disposal. This appears prescient to me.

Overall, my prediction for the digital future is "not much change". That might strike many as ludicrous. We are told daily that the world is changing at an ever-faster pace. Some predictions for 30 years hence talk about transmitting emotional states and having buildings with living fabrics. Autonomous cars and AI are just around the corner it appears. 5G will appear shortly when we are promised a step-change in all we do. "Not much change" flies in the face of conventional wisdom, of recent experience, of what we hear daily.

I hope that in reading this book you have understood why I have come to these conclusions. I hope that my track record of forecasting in this area gives you some confidence that my views might have merit.

With that, here are my views on what the world will be like in 2027, 2037 and 2047.

11.2 The world in 2027

Before looking forward ten years, I look back ten, to remind us as to how much can change in a decade. 2007 was a pivotal year - it saw the introduction of the first iPhone. That has led to a huge revolution, making data connectivity on the phone valuable and leading to endless apps. It gave a huge boost to Twitter (created in 2006) and eventually to Facebook and YouTube as data networks improved. It allowed many of the concepts foreseen by me and others such as using a phone as a boarding pass and finding a nearby Starbucks. Software in the phones improved massively including intelligent agents such as Siri, making many tasks much simpler. It has changed behaviours significantly, for example on trains most used to read newspapers, now most look at their phones. Indeed, many continue to look at their phones while walking down the street and when in the toilet! Of course, we had mobile data connectivity 20 years ago, but the last 10 has made it useful and accessible.

Home broadband connectivity also improved massively, from a stage where only the very lucky few had a Mbit/s to one where 50Mbits/s is quite normal, and some can go much higher. Home Wi-Fi is much better - faster, more reliable, able to connect more devices and with better range. But the home itself has not changed much, other than connected HVAC controllers for some. Nor has transport other than the ability to stay connected on trains and planes.

Computers, laptops and the Internet have remained very similar both in form-function and in capabilities. Most of our jobs have remained the same and are conducted in the same manner (still too many meetings).

So, to many, a decade of a lot of change, but primarily triggered by one event - the ability to use data effectively on a mobile device. Is there another such event coming in the next decade?

I believe there is a similar event - the ability to simply and cheaply connect IoT devices wherever they are. But this will be much less noticeable, leading to a slow improvement in reliability, functionality and productivity. Unlike the iPhone that was everywhere within a year, the IoT will take more than a decade to gain widespread penetration.

Looking at the bigger picture of enablers and developments, 2027 will see:

- IoT connectivity becoming more prevalent but primarily in business activities such as manufacturing and agriculture, resulting in improved productivity but little change in daily lives.
- AI will become very good at specific tasks. Areas like language translation and speech recognition will be excellent, and the "virtual assistant" function that devices are starting to provide will be very

effective. Adverts will become highly targeted online. But AI will not have transformed the world.
- Connectivity will appear to be perfect. Mobile and broadband networks will provide much greater speeds than we need, and the most prevalent not-spots such as on trains and in buildings will be mostly filled in, often with Wi-Fi. But, apart from coverage, networks are already often at the point of appearing perfect, so this is not a major change.

For the individual the virtual assistant will be much more powerful, suggesting entertainment, filtering and in some cases auto-responding to messages, planning travel based in calendar appointments, providing personalised news feeds and more. This will not enable anything new that was not previously possible, but will make some tasks much simpler, or indeed automatic, such as finding the best insurance deal when due for renewal. However, the handset, and indeed other communications devices, will remain much the same, basically consisting of a touch-screen in an outer case.

Homes have not changed much in the last 10 years, and are unlikely to do so in the next ten. My predictions are:

- Automated home HVAC systems will become ubiquitous during which time they will also gain the intelligence to adapt automatically to the diaries of the home occupants.
- Smart speakers such as Amazon's Echo, will be widely deployed within five years, but mostly used as a kitchen radio.
- Most innovation will occur at the device and the individual level rather than at the home appliance and household level.

The office potentially will see much more digital because it makes commercial sense. However, we might not notice since the aim is not to add new functionality but to reduce maintenance costs. Specifically:

- IoT will progressive be deployed in lights, plants, waste bins, coffee machines and more, starting within 5 years and throughout the coming decade.

- Biometrics will be progressively used to open doors and enable IT systems, starting almost immediately completing by 2027.
- Robotics will make some impact, with robotic vacuum cleaners and possibly other solutions to tasks that need to be performed daily. This will only just be starting to occur by the end of our time period.
- IT in the office will nearly all go away as BYOD predominates. This is already happening and will complete within 5 years. Even conference room IT systems will broadly disappear with simple TVs/projection units left for slideshows.
- Of the major employment sectors agriculture will see some steady productivity improvements continuing current trends, vehicle maintenance will grow, retail will continue to decline at approximately its current rate, construction and hospitality will be largely unaffected, and manufacturing will see ever-growing penetration of robots on the production floor.

Regarding transport my prediction is very little change. There will be better on-board connectivity – we are already well on the way to achieving this and it will broadly complete in the next 5-10 years. We will have simplified ticketing and payment – again this is already in place on some transport systems (eg buses and underground trains in the UK) so extending it to other areas is a relatively minor task.

For leisure I predicted that:

- Each form of leisure will get its own specialised digital enhancement from the equipment used through to the supporting websites and apps. This will burgeon over the next decade.
- Digital entertainment such as games will continue their steady growth towards ever-greater realism and complexity, adding voice control and other AI-related features.
- New forms of digital entertainment, enabled by AR/VR and similar will be trialled and adopted by a few, but will be struggling to find their first mass-market application.

Predictions

So broadly, the largest changes will be in the virtual assistant, at work, in various work environments and in our leisure activity. Other areas might see little change.

The next decade will be a disappointment for many digital companies. 5G will be introduced but prove to be of little interest and end up as merely a minor enhancement to 4G. FTTH deployments will mostly be shelved as global experience shows no real interest in data rates beyond those available over fibre-to-the curb and other similar solutions.

Digital companies will find it a hard time, specifically:

- Apple with lose their sparkle as phone and laptop replacement cycles extend and there is ever-declining reason to upgrade to the latest iPhone.
- Handset manufacturers in China and Asia-Pacific will tend to dominate as they have in almost all other consumer products.
- Major telecommunication equipment vendors such as Nokia and Ericsson will suffer as infrastructure deployments stall.
- Mobile operators will gradually tend towards wholesale providers as mobile virtual network operators (MVNOs) blossom. These MVNOs will adopt many different tactics with some replicating Google's Project-Fi model where Wi-Fi plays a major role in connectivity.
- For fixed operators it will be business as usual.
- The dominant Internet companies - Google, Amazon, Facebook, will continue to dominate, but growth rates will gradually fall as it becomes harder to find new innovations that can be monetised.
- The companies with new business models - Uber, Airbnb and others, will remain but will struggle with employment laws, data privacy requirements, some societal discomfort and so on. Finding a way through all of this will take time and sap management attention.
- New companies will emerge, of course. But the rate of emergence will slow as new ideas are harder to come by and VCs perceive that the returns in digital are falling.

This will be unfolding against a background of volatile social and political issues. This book is not about what will happen in politics, but it seems likely we are set for a turbulent time, whatever the outcome. This might dampen economic growth and investment certainty, making it even less likely that large-scale new digital infrastructures or projects will be funded.

Some of this turbulence is caused by digital and we may need a period of less digital change to enable society and its structures to catch up. While politicians do not control digital change, societal push-back in various ways might make digital less welcome. Unfortunately, at the moment, the links between those evolving digital and those evolving societies are very weak so this may not happen in any coordinated manner.

11.3 The world in 2037

As before, I will start by looking back over the same time-period as I am forecasting, in this case 20 years to 1997. This was just about the time that the Internet became mainstream. Many will have gone on-line for the first time around then, using dial-up modems and trying to understand what the Internet was all about. The dot.com bubble was about to inflate as Amazon and many other new companies emerged. By 2001 the bubble had burst, leaving most to go out of business but allowing the strongest to solidify their leading position. Mobile phones were widespread but rarely used for data. Laptops were ubiquitous but tablets still to be invented.

Like our look at 2007 - 2017, the period 1997 - 2007 had one major new concept, in this case the Internet. Its impact was massive and was at the heart of almost all digital changes since. Can we expect another new concept, as powerful as the Internet, to emerge in the next 20 years? My view is probably not. The Internet was a one-off rather than a part of a larger trend of new concepts coming along every decade. In the same way that it is only possible to introduce electricity once, it is only possible to introduce computer connectivity once. However, AI may well become as mainstream as the Internet during this time, and while not having quite as much impact as the Internet, it could still be very significant.

Looking at the bigger picture of enablers and developments, 2037 will see:

Predictions

- IoT connectivity become very widespread, especially in all branches of business. We will be surprised when devices are not connected and annoyed when they cannot act intelligently and call for help when they need it.
- AI will have become extremely good at specific tasks. The challenges of more general AI will be well understood and research underway to address them. Big datasets will provide useful training material for machine learning systems.

For the individual their virtual assistant will now link to the IoT world, interacting with items around the user and changing their behaviour as needed. Much of this will happen without individuals even being aware of it.

Around 20 years from now AI will have evolved to deliver some useful in-home features that are not currently clear to us today, perhaps coupled with greater IoT connectivity in the home.

The office will continue its evolution. Specifically:

- IoT deployments in lights, plants, dustbins, coffee machines and more, which started to happen in the previous ten years, will become widespread in this period.
- Robotics will make an increasing impact, with robotic vacuum cleaners and possibly other solutions to tasks that need to be performed daily.
- Of the major employment sectors vehicle maintenance will decline rapidly as electric vehicles predominate, retail will be almost entirely on-line, but construction and hospitality will remain largely unaffected. Robotics in manufacturing will be widespread.

Digital will have gradually transformed public services, bringing them up towards the level of digital in the private sector. Healthcare will see major progress as AI becomes more widely accepted as a valid diagnostic tool.

Productivity might now be improving, using AI, robotics and IoT to deliver more crops, better products and reduced maintenance. This will lead to some job losses and society will be debating whether to push for continued full employment or to make shorter working weeks the norm. Debates on paying a basic income to all will be widespread.

We might hope that the political turbulence will settled down in the 10-20 year time horizon, replaced by debates about the balance of work and leisure and what to do with more leisure time. Digital will help by expanding the scope of leisure and enabling favourite pastimes to take more time - such as playing at being a virtual football manager when the football team is not in action.

11.4 The world in 2047

As before, I start by looking back over the same time period, now 30 years to 1987. This also saw an era with one big invention - in this case the mobile phone. While there had been car-phones for the rich, the late 1980s saw the first hand-portable phones. This led to phenomenal growth in subscriber numbers during the decade. Miniaturisation made it a decade of much change for electronics. Laptops became possible and portable. Pagers were widespread. Camcorders became widely available and digital cameras were just around the corner (all to eventually be subsumed into the smartphone).

It really does appear that we get one major digital breakthrough a decade – the mobile phone 30 years ago, the Internet 20 years ago and the iPhone 10 years ago. I suspect this is more coincidence than any "law" but it does suggest that predictions should allow for something happening every decade. My predictions are for IoT to be the breakthrough of the next ten years, AI for the decade after that and perhaps robotics (using AI to gain intelligence) for the third decade.

Not many of my predictions from earlier chapters were for the period 20-30 years out. As always that is partly because it is harder to imagine what we might want and what we might invent that far into the future. Inventions such as the mobile phone in the 1980s led to the iPhone in 2007 and to virtual assistants and electronic payment by 2017. Predicting the iPhone would have been much harder if the mobile phone had not yet proven possible.

Thirty years out marks a point where many believe a "singularity" will occur. This is where AI becomes more intelligent than humans and then rapidly takes off; inventing its own ever-more superior machines in a spiral that quickly goes beyond our control. I am sceptical (after all, this had been forecast to happen in 2001) but nevertheless there is much uncertainty about AI and the role it might play by this point. In the period between 2037 and 2047 it is likely that this uncertainty will be resolved and either AI will remain a useful tool in specific cases (eg playing Go) or it will emerge as a powerful addition to humankind. In the latter case, predictions as to what might happen are nearly impossible from this far off.

For these reasons, I do not have any specific predictions 30 years out. If pushed I would tend towards the same predictions as 20 years out, but with high uncertainty that something important will have been missed.

11.5 National differences

Will the digital world of the future be different according to where you live? For example, it could be imagined that China will follow a different path to suit its political needs or that Japan will adopt robots for the care of the elderly to a far greater extent than other nations.

But differences to date have been very small. Most digital services are inherently global and are adopted everywhere. China does have its own versions of search engines and social media websites, but these are essentially the same as available from Google and Facebook. There is nothing in my predictions of the future to suggest that this will change.

Of course, the speed of adoption will differ. Richer countries will make use of IoT, AI and robotics before poorer countries. In some places cultural barriers will slow particular concepts. Robots may need to look different in some cultures. Overall, one of the roles of digital has been to make the world a smaller place. In doing so it has reduced the likelihood of national differences.

11.6 Why so pessimistic?

As I have noted before, many will see these predictions as hugely pessimistic. The Internet abounds with those who predict so much more. For example, Ian Pearson, "a futurist with an 85% accuracy record", came up with a list[24] in 2015 which I have summarised below. (Where the predictions cover areas I have discussed at length I have provided more detail).

1. We could start seeing delivery drones finally start making deliveries in the next two years.
2. A Hyperloop could take us in between cities in just six years.
3. Machines could start thinking like humans as early as 2025.
4. Space trips designed to send people to Mars could start taking place in 2030.
5. Prosthetics could get so advanced in the next 10 years they could give people new skills.
6. Clothing could give people superhuman skills in the next 10 years.
7. Virtual reality could replace textbooks during the next decade.
8. The smartphone will become obsolete by 2025 thanks to advancements in augmented reality. It will be possible to pull up screens in AR via a tiny bracelet or other piece of jewellery in the next 10 years, making it unnecessary to carry around a smartphone.
9. Self-driving vehicles could be ubiquitous in the next 10 years.
10. 3D-printing could be used to construct more houses in 20 years.
11. People could start using robots to do work around their house and provide companionship starting in 2030.
12. We could live in a Matrix-like virtual world by 2045.
13. People could also become Cyborgs by 2045.
14. People could control their home settings using artificial intelligence by 2040 as well. By 2040, AI will be built into buildings themselves, so you can talk to the building and ask for adjustments in temperature or lighting.
15. Super tall buildings could function like mini-cities in the next 25 years.
16. We could rely entirely on renewable energy by the year 2050.
17. Space tourism could be feasible in 2050, but likely only for the very wealthy.

---

[24] See http://uk.businessinsider.com/ian-pearson-predictions-about-the-world-in-2050-2016-7/?r=US&IR=T/

It is hard to know where to start here. A few of these appear plausible such as being able to talk to a building and ask it to adjust its temperature. But most seem incredibly optimistic – such as AI being able to think like a human by 2025 (unless that human is a five-year-old). Take, for example, the second one - that a hyperloop will take us between cities by 2021. At the time of writing in 2017 small sections of hyperloop were being tested in the desert. To connect two cities would require (1) the technology to be proven (2) a tunnel to be dug, or a tube to be laid, between the major cities (3) trains to be mass produced and tested. Given that it takes more than 20 years for new train lines to be laid into major cities and similar lengths of time for major tunnelling projects, it is utterly impossible that this could happen by 2021. Even 2041 is a stretch. And that ignores the question of economics (recall that major projects like the Channel Tunnel went into bankruptcy). Even more obviously wrong is the first - that drone deliveries will start in 2017. It is now 2017 and they haven't.

All I can say is that if you prefer this set of predictions to mine then find a different book!

It all comes back to Peter Thiel and "we wanted flying cars and we got 140 characters". We want an exciting future where the world is a visibly better place and we like to be told that this will occur by visionaries. We often get something much less.

11.7 In summary

The world has seen one major "enabler" per decade in terms of a new concept becoming mainstream in the decade of the last thirty years. This started with mobile phones, then the Internet, then the iPhone/touch screen devices. There is no particular reason why this pattern should continue but I predict that it might. Over the next thirty years I anticipate IoT, then AI, then robotics as being key enablers. However, the big difference of these enablers is that they do not directly and immediately provide a change for the consumer, unlike the mobile phone. Instead they will enable better productivity, more reliable and responsive products and faster innovation. While these are all good things, there may be downsides in reducing employment and risks of unintended consequences from AI.

Figure 11-1 provides a summary on one picture.

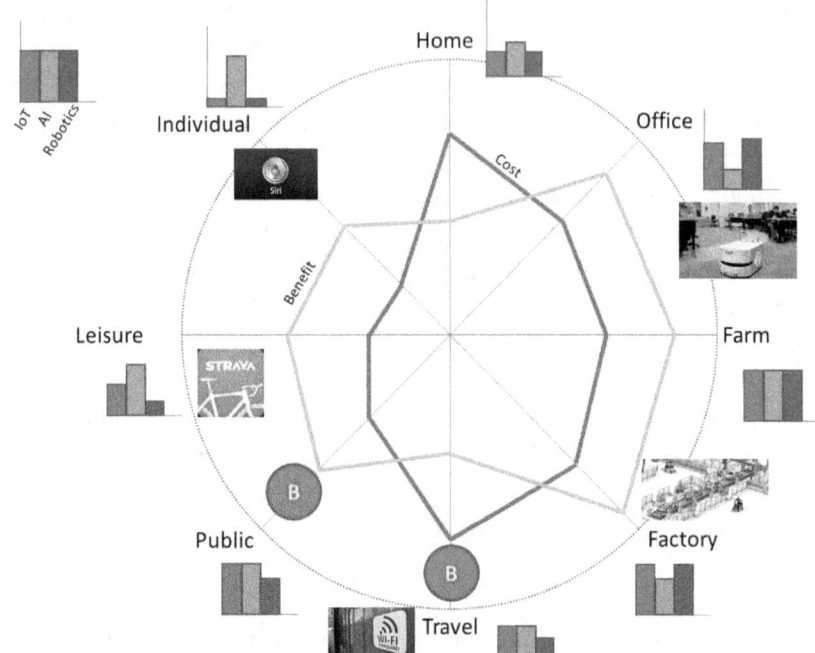

Figure 11-1 - A summary

In this figure, eight different environments are grouped around the circle. For each environment there is a cost and a benefit of new digital concepts. Where the benefits are greater than the costs then implementation should follow, but in some areas there are other blockers, indicated by the "B". For each area the relative importance of the three enablers is shown in bar charts, and where relevant a graphic represents the key development in that sector.

The next chapter looks at what these predictions would mean for companies in the digital space.

12 Structure of the digital industry

12.1 Introduction

In this chapter I look at what might happen to the companies and organisations in the digital space.

12.2 Our current behemoths

At the time of writing the world's most valuable company is Apple. Tesla is worth more than Ford, despite selling less than 1% of the number of cars. Google, Amazon and Facebook are seen as some of the most powerful companies in the world. Most of these companies are barely 20 years old. Figure 12-1 shows the world's top ten companies by value.

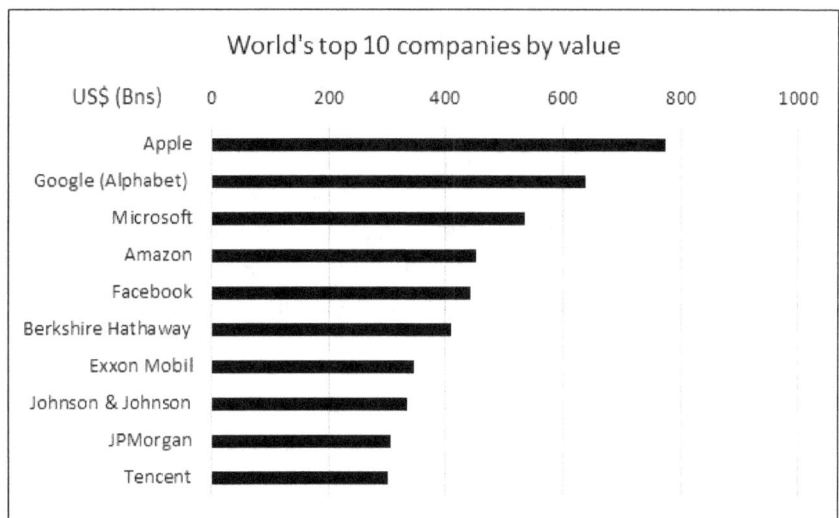

Figure 12-1 - World's top companies by value

A common characteristic is that all the top five are American, founded on digital and somewhat monopolistic. Each dominates its particular space and often has "network benefits" that make entry by another player very difficult. Some are acquisitive, buying up companies with promising new ideas that might, perhaps, become competitors. Tencent is also a digital company, but based in China.

New companies do emerge, and some such as Uber are discussed below. The start-up scene remains reasonably vibrant. But the future, at least for the next few years, does seem to depend heavily on what these behemoths decide. Happily for us, they often decide to innovate and to fund new ideas even where the profitability is uncertain.

What about the longer term? My predictions for the future suggest that the need for the services offered by these companies will remain. Most obviously we will still need Internet search and still want to use social media. But the way that we do these things will change, and the existing players will need to adapt. They may also be buffeted by social and Governmental forces. By way of example, I take a look at what the future may hold for each of Apple, Google, Amazon and Facebook.

Apple. Apple's success is predominantly due to hardware. Their revenue comes mostly from sales of iPhones, iPads, laptops and other devices. Their growth has been down to introducing new concepts such as the Apple Watch. At the time of writing, new concepts were increasingly rare. New products were proving hard to devise and updated versions of the iPhone had less and less novelty. Consumers were choosing to wait longer to upgrade their device. Continuing to grow would seem hard. Equally, Apple users are loyal and are locked into the Apple eco-system, making a decision to move to a different supplier hard. We will continue to need phones, tablets and laptops for the foreseeable future. Apple may find profit margins slowly falling, but their future does not appear in much doubt.

Google. From an initial business model of providing search, Google has proven itself very adaptive. Its acquisition of Android and the subsequent free availability of the operating system has given it a key role in the mobile space. It has a wide range of solutions from Google Earth to Google Now and is famous for experimenting with more extreme ideas such as Project Loon, which provides mobile coverage from high-altitude balloons. It has been a pioneer in driverless cars and is threatening to revolutionise the world of mobile telephony with its Project-Fi service. Is there nothing Google cannot turn its hand to?

Structure of the digital industry

The ride has not been completely smooth. Litigation against Google is growing for a range of reasons from monopolistic positioning to (fully legal) avoidance of local tax payments in some countries. Many of Google's services are blocked in China.

Google is investing in the future envisaged here. In 2013 it acquired Boston Dynamics, a key pioneer of robots and more recently it acquired a key AI player - DeepMinds. It is entering the home with its own smart speaker. It has proven itself very adaptable, changing direction when appropriate. All of this is helped by a healthy profit from the core business, allowing for experimentation, long product gestation periods and acquisitions whenever needed. Google also promotes a very supportive working environment, attracting many of the best minds.

As a result, Google appears very well placed. Its core business looks stable and it appears to be invested in most of the key trends of the future. It is not placing bets in areas I would regard as unlikely to succeed. Of course, the same could be said for Nokia when it held over 45% of the global handset market, but in Google's case its near-monopoly on Internet search means it is much harder for its core business to come under attack. Various actions from Governments and regulators appear inevitable, but as we have seen with Microsoft over the years, these are typically little more than an annoyance.

Amazon. Is there nothing that Amazon does not sell? From relatively humble beginnings as the Internet's on-line book store, Amazon has expanded into all aspects of our lives. Like Google it has shown itself willing to experiment. Some have gone well - with Kindle it has a near-monopoly on e-readers. Others less well - its tablets have never really caught on. Not only is it a retailer but also a key provider of cloud computing services. Along with Microsoft it dominates the cloud space, and Amazon Web Services is the place most start-ups go for storage and computing power. This leaves it relatively well placed for the world of AI where it can combine the vast datasets it collects along with vast processing power. Indeed, its "recommends" service is an early example of the application of AI to retailing.

It is also strong in robotics, pioneering robots in its warehouses and drones as a delivery mechanism. The former sounds sensible, the latter less so. Like Google it is making a play for the connected home with Amazon Echo, and seems, at the time of writing, to have the upper hand.

Just as with Google, the future seems secure. It has the same near-monopoly on generic Internet retail and the same willingness to experiment and adapt. It has similar exposure to the key enablers of the future and no competitor in sight.

Facebook. Social media seems a little less secure than search and retail. It has often proven elusive to find a revenue stream, and some like Twitter still struggle. Facebook has now found an advertising model that works well for it, but the longevity of this is unclear.

Competition is more likely than for the others. New concepts abound due to the relative ease of starting them such as Instagram. Sites for specific purposes or geographies are ever-more used including LinkedIn for professionals and China-specific sites such as Baidu. I predict many more leisure-specific sites such as Strava for runners and cyclists. While many will continue to use Facebook alongside these, the time spent on Facebook may diminish.

Facebook struggles to innovate in the same way as Google and Amazon - its passage to a broader product base is far less clear. Ventures into new forms of mobile communications, for example, have been ill-founded and generally failed. It owns Oculus Rift, a leading provider of VR equipment, but Oculus has yet to make an impact and faces growing competition from AR and other providers. As discussed earlier, I believe AR is much more important than VR.

Facebook is also the subject of ever-growing regulatory and Governmental scrutiny, more so than the other companies listed here. This is because it is seen as having increasing influence on "news" and on public opinion. Some believe it impacted the 2017 US Presidential election. It may need to change its policies for content checking, potentially at great cost. Indeed, there is some doubt as to whether the current social media model can survive the issues and challenges it is now seeing.

For all these reasons, I would not be surprised if Facebook no longer existed 20 years hence, although there will be other social media platforms to take its place.

In summary, the advent of digital has enabled some of the world's largest companies. Most of these are well placed to continue as leaders into the future envisaged here. But will they be joined by others?

12.3 The rise of the Ubers

It is often noted that the world's biggest taxi company has no taxis, the world's biggest hotelier owns no rooms and the world's biggest retailer has no shops. (Uber, Airbnb and Amazon, respectively.) Uber and Airbnb are pioneering a new "sharing" model where they facilitate those with cars or rooms linking up with those who want rides or somewhere to stay.

For the customer, Uber changes only the convenience. It is still a taxi service in a car with a driver, taking them from one place to another, along the same roads. Airbnb does change things a little by offering accommodation in a new type of room – often in someone's home. Conversely, for those in the industry it can change everything. Licensed taxi drivers are suddenly finding that customers are going elsewhere and they are unable to compete. Uber drivers are winning business but only by being self-employed, without the safety nets that come with employment contracts. All of which places more pressure on society to be there to pick up the pieces if things go wrong.

Interestingly, while many see Uber as a sign of just how quickly things are changing, they do not feature on most predictions of the future, nor need they. Previous predictions often talked about automatically calling a taxi and automatically paying for it. They did not feel the need to discuss whether it was a licensed taxi or driven by a self-employed agent of an Internet entity. Indeed, predicting which companies will survive, which will fail and where new entrants will make their mark is particularly hard because it can often depend on good or bad strategic decisions from a few key individuals. But unless you are employed by these companies or an investor in them, it does not matter overly.

So while Uber is a feature of many discussions and seen by many as a symptom of how the world is rapidly changing, in practice it may not be particularly relevant for most aspects of future prediction other than, strangely, employment law.

Will Uber be joined by others? Many entrepreneurs are now looking for other areas where the same sort of disruptive business model might work. Emerging companies include Deliveroo for food deliveries using cyclists. Real estate feels like an area due for a shake-up but has proven remarkably resilient in the past. Amazon's Mechanical Turk is innovative but has not revolutionised working culture. But just as few saw Uber coming, it is likely that the next Uber will catch us by surprise, but seem obvious with hindsight.

My best guess is that many have given this much thought, many ideas have been tried and all the low-hanging fruit has been picked. Smaller successes might follow but little on the scale of Uber.

12.4 The digital enablers

The digital world is a broad one and alongside these prominent companies there are thousands of others, employing millions of people. What of them?

12.4.1 Connectivity

Connectivity is a huge space. Most countries have a fixed telecoms provider, some have cable operators. Most have competing cellular operators, typically three or four. Some have MVNOs including Wi-Fi connectivity providers.

I have said little about connectivity other than it is already more than good enough for most and that little change is expected. While there is much hype around Gbit connectivity, in practice few value this, and so operators will not spend much to deploy it.

Connectivity is already well on the way to being a utility, with little revenue growth, and increasingly little innovation. Its future is safe since we will always need connectivity and always see it as sufficiently critical to pay whatever price is asked. But a mix of competition and regulation means that returns on investment will converge on those typical for utilities. Companies like Vodafone will be seen as stable, zero growth entities that pay good dividends.

Structure of the digital industry

As I set out at length in "The 5G Myth", the business-to-consumer (B2C) role of these companies may diminish. Companies like Google may offer Project-Fi type connectivity, forming a direct customer relationship. Mobile operators may tend towards wholesale business-to-business (B2B) provision as a result.

There are some large companies who specialise in providing equipment and services to the connectivity providers including Ericsson, Alcatel and Huawei. As networks stabilise and investment drops to maintenance levels, these companies will see falling sales and a shift in focus away from hardware and towards software. Indeed, some already have, and are shrinking as a result. Expect some companies to fail, some to be split apart and the more attractive elements sold, and some to enter a lower-profitability era.

In summary, connectivity providers will generally be stable, zero-growth utilities, while their suppliers will go through a turbulent period of falling sales.

12.4.2 The Internet

By "the Internet" I mean everything that sits between the connectivity providers and the websites or apps. Some of this is hardware such as the routers provided by Cisco. Some is generic software running databases such as Oracle and SAP. Some is a mix of these coupled with integration skills such as IBM. Cloud services also fall into this category, such as Amazon's AWS and Microsoft's Azure.

Much of what has been said about connectivity providers also applies here. The core Internet "plumbing" is necessary, growing steadily, and will continue to be needed in whatever future transpires. For example, the router market will remain important for the foreseeable future. However, unlike connectivity, competition can come from elsewhere, such as Chinese and Asia-Pacific companies. Huawei has seen meteoric rise and as this market becomes more of a commodity, and as the Chinese market grows, then the risk of price-competition becomes ever greater.

The cloud services market is a growing one. But it is also a commodity and hence one where margins will remain tight.

The consultancy and systems integration market in this space is vibrant, populated by players such as Accenture. Consultancy thrives on change, but if connectivity and the core Internet become more stable, the need for their services might fall (but equally other areas may open up such as IoT integration into businesses).

In summary, another future-proof and stable market, but another one where growth will be limited and profitability no better than industry averages.

12.4.3 IoT

The IoT market is currently small. There are few large players involved. Some mobile operators are seeking to provide connectivity, but so are start-up companies like Sigfox. Equipment is generally made by small players in small quantities, often using general-purpose chipsets or platforms to keep costs down.

The larger manufacturers and Internet players such as Amazon, Apple and Google have made plays for the home IoT market with smart speakers and home eco-systems but it feels more like a strategy to protect their core business rather than because they believe it is a market worth attacking in its own right. (And, as I predict, the home IoT space will be small.)

Overall, the IoT market is likely to be small relative to, eg, the cellular market, and highly fragmented into myriad different applications, industries and requirement sets. It seems highly unlikely it will foster another Apple. I anticipate that:

- Devices will use custom chipsets, delivered by existing players such as TI, Analog and SiLabs. Volumes will be high but margins thin, akin to the Bluetooth chipset market.
- Connectivity will be provided by a mix of mobile operators, new emerging operators and self-provision. The low connectivity value and split nature of provision will mean such companies will be small but profitable.

Structure of the digital industry

- Internet platforms will come from entities such as Cisco (eg who bought Jasper, a specialist IoT platform provider) and some emerging companies.
- Each vertical, such as agriculture, will have its specialist solutions provider, delivering complete packages and offering data analytics.

In summary, I expect IoT be a growth area engendering much innovation. It will lead to many new companies, few of which will be particularly large. It will also deliver a small amount of growth to larger players such as Internet companies and mobile operators.

12.4.4 Devices and software

The category includes handsets, tablets, computers, smart watches, etc. It also covers both the hardware and the software including operating systems and basic software such as Microsoft's Windows and Office. It is a huge eco-system and includes companies like ARM which provide the core processing templates.

While the need for devices is not going away, nor changing materially, neither is it growing. Indeed, there is evidence that replacement cycles are lengthening as it gets harder to differentiate new products. I touched on this in the discussion above on Apple. As innovation slows, the market commoditises, with reduced profitability and a tendency for manufacturing to move to the Far East.

The handset market has seen much change over the years, with the exit of some key brands such as Nokia and Motorola and the growth of new entrants from the Far East. The laptop market has been similar, with players like IBM selling their branding and expertise to Lenovo. Google has bought, and sold, on a number of occasions, most recently buying around 2,000 engineers from HTC.

In summary, yet again this looks like a relatively stable market, with commoditisation taking its toll on those with a higher cost base.

12.4.5 AI

There are few pure AI companies, and those that do emerge, such as DeepMinds, tend to rapidly get acquired by larger players. It is hard to sell "generic AI" although there may be some value in machine learning packages or languages. Instead, the value of AI mostly comes in its application. Siri is a better virtual assistant as a result of AI and this allows Apple to sell more iPhones.

Hence, I would expect existing companies to benefit from AI. Those with the most to gain are the ones that have large datasets, or access to this data, and where greater intelligence can lead to a better product. This will be true of a plethora of companies from Amazon to medical suppliers and from call centres to virtual sports coaches. A few start-ups will make their founders very rich on the road to acquisition but this will tail off as the hype fades.

12.4.6 Robotics

At present, some of the world's largest robotics companies are entities such as Bosch and ABB that have an extended manufacturing heritage. There are many start-ups in most "top 10s" such as Boston Dynamics. Companies such as Dyson make robotic vacuum cleaners.

My expectation is that most robots will be specific to certain tasks - such as robotic vacuum cleaners, or production line welding robots. It may be that companies that make vacuum cleaners are best at making robotic vacuum cleaners and companies that make industrial manufacturing equipment are best at making manufacturing robots. Some specialised entities might provide platforms on which more specific designs can be based.

Robotics looks like a long-haul, with much research and development needed and sales growth being slow. This would bias it towards the large, well-established industrial companies with a corporate ownership structure that allows for long-term investment - perhaps companies like ABB.

It will take many years, perhaps decades, before it is possible to predict with confidence who the winners and losers in the robotic era will be.

12.4.7 OTT and apps

There is an app for everything it seems. Occasionally a new app "goes viral" such as Angry Birds and more recently Pokémon Go, although such apps tend to

Structure of the digital industry

fade away after an initial explosion. Some apps are from new entities for who the app is their only business, such as WhatsApp. Others are simply part of a broader service such as apps provided by banks or airlines.

This is a world that works well, that we understand and that allows for innovation. There seems no obvious reason for it to change in the future I envisage.

12.4.8 Digital content

Huge value now resides in content. So-called "creative industries" make up ever-larger fractions of GDP. For the most part there is little reason to expect this to change. New content may appear for VR consumption, but I expect this to be relatively small.

12.5 The winners and losers

The good news for most companies in this space is that the pace of change is unlikely to be such that they will be swept away, as happened with the Internet revolution. The bad news is that with little change there will be limited growth and more commoditisation, reducing profitability.

The major winners look to be those who are already winners - Amazon, Google, Microsoft and similar. The losers may be the equipment suppliers who will both see a falling market and are most exposed to competition from other countries. More generally, if IoT improves product reliability then product lifecycles may extend, reducing the market size for companies making a wide range of products from phones to dishwashers to street lights.

The biggest question marks are around companies who impact politics such as social media players. They may see significant new regulation as well as other threats to their business model.

13 The future on two pages

Peter Theil will continue to be disappointed. We may have wanted flying cars, living buildings, holographic displays, and more; we will continue to get 140 characters, Siri and Alexa.

The key enablers of change will continue to be the Internet, ubiquitous connectivity and flexible touch-screen devices. In addition, IoT will enable much in the next decade, AI will really come to the fore in the decade following and robotics might become commonplace the decade after that. Because we have already picked the low-hanging fruit from the existing enablers, it is these new enablers that will drive change in the digital world.

Most of these enablers are associated with business rather than the individual. Hence, the change noticed by the individual may be relatively small compared to the change of the last 30 years. Individuals will see an ever-better virtual assistant functionality from their devices as solutions such as Siri steadily improve using emerging AI techniques. In the home some new connected devices such as smart speakers and home IoT products will be installed but home automation will not be widely deployed. Leisure interests will expand, with each genre (eg cycling) gaining apps, on-line communities, additional functionality and where appropriate monitoring from IoT devices. This will allow us to spend more time on our favourite pastimes, as indeed we may need to if enhanced productivity and automation leads to fewer jobs.

In business, the office will see widespread deployment of IoT, biometrics and robotics, mostly as a way to save costs on administrative and maintenance staff. Some sectors will make extensive use of IoT to improve productivity such as agriculture and manufacturing. Some will decline further due to changing habits such as retail. Some will be broadly unaffected such as construction and hospitality. Vehicle maintenance, which is currently a huge employer, may decline as more electric vehicles are introduced and as car sharing gains traction.

Transport will not change materially other than we will be better connected while travelling, have more journey information and see a gradual growth in driverless vehicles (cars, trains, buses, etc).

The future on two pages

Society may become ever-more concerned about the changes wrought by digital, and there may be some push-back. Contract law will adapt to, and will change, the new "zero-hours" approach to employment. Social media will be charged with cleaning up undesirable content and controlling fake news. Autonomous cars, robots that provide companionship to the elderly and similar developments will raise difficult ethical questions. Privacy and security concerns will limit the scope of big data and AI in some areas, and may slow the introduction of IoT.

Today's large digital companies such as Google and Amazon will continue to dominate well into the future. New players such as Uber and Tesla will emerge but at a slowing rate. A few, such as Facebook, might struggle as regulation bites. Connectivity providers such as mobile operators will become utility-like, and their manufacturers will struggle.

In essence, the key gains will be in convenience, productivity and reliability. The world will be a similar place to today, but will work better.

This will strike many as pessimistic when others talk of flying cars, cyborgs and AI that is superior to humans. I would suggest it is pragmatic realism.

Index

3DTV, 39
5G, 35, 36, 37, 46, 92, 96, 110
A380, 71
ABB, 113
Accenture, 111
agriculture, 64, 112
AI, 40, 59, 81, 90
AIDS, 5
Airbnb, 108
Alcatel, 110
Amazon, 5, 44, 59, 60, 94, 96, 97, 104, 105, 106, 107, 108, 109, 110, 111, 113, 114
Analog, 111
androids, 18
Angry Birds, 113
Apple, 11, 18, 40, 42, 63, 96, 104, 105, 111, 112, 113
AR, 35, 39, 40, 44, 50, 101, 107
Arpanet, 36
ARPU, 6
autonomous cars, 73
autonomous vehicles, 45
babel fish, 20, 21
Baidu, 107
battery, 44
big data, 42, 91
Bill Gates, 27, 29
Bluetooth, 7, 8, 9, 63, 72, 73, 76, 111
boarding passes, 72
Boeing 737, 71

Bosch, 113
Boston Dynamics, 113
Brexit, 87
broomsticks, 17
bullet trains, 72
BYOD, 62, 70, 72, 95
cancer, 5, 6, 81
Cisco, 110, 112
climate change, 5
cloaking device, 18
cloud, 110
Concorde, 71
congestion, 74
construction, 67
contactless, 73
Cooper's law, 32
cryptography, 49
cycling, 76, 77, 78
DeepMinds, 106, 112
deluminator, 17
denial of service, 49
dermal regenerator, 18
DNA, 5, 91
Donald Trump, 87
dot.com, 97
drones, 64
Echo, 44, 59, 60, 94, 107
education, 84
electric vehicles, 4, 44, 66, 98, 115
Elon Musk, 74
Ericsson, 110
Ethernet, 55

Index

fake news, 5, 87
fingerprint
 quantum, 49
floo powder, 17, 19
flying cars, 15, 22, 34
foldable, 22, 26
FTTH, 36, 37, 96
GDP, 6, 114
GDPR, 88, 91
Gilder's Law, 32
Glonass, 76
Google, 7, 21, 39, 40, 41, 42, 45, 47, 71, 87, 96, 104, 105, 106, 107, 110, 111, 112, 114
GPS, 76
HAL, 20
happiness, 6
hard disk, 32
Harry Potter, 17
healthcare, 81
Henry Ford, 16
Hitchhiker's Guide, 19
holodeck, 18
holographic displays, 24, 26, 58, 115
home entertainment, 57
hospitality, 68
hotels, 10, 67
howler, 17
Huawei, 110, 121
HVAC, 55
hypospray, 18
Ian Pearson, 100
IBM, 20, 27, 28, 81, 110, 112
Industry X.0, 68
innovator's dilemma, 28

Instagram, 107
Internet, 35
invisibility, 25
invisibility cloak, 17
IoT, 38, 59, 62, 111
iPhone, 3, 13, 16, 92, 93, 96, 99, 102, 105
IPv4, 35
IPv6, 35
IQ, 42
jetpacks, 22
jobs, 88
Jules Verne, 16
Keynes, 90
lawn mowers, 23
lawnmower, 8
leisure, 76
LinkedIn, 107
Lord of the Rings, 17
magnetic levitation, 72
manufacturing, 68
marauders map, 17
Mechanical Turk, 109
Microsoft, 19, 42, 50, 58, 106, 110, 112, 114
Minority Report, 19, 25, 61
MOOCs, 83, 84
Moore's Law, 32
Motorola, 16, 112, 121
MRI, 21
MVNOs, 96, 109
nanoprobes, 18
Netflix, 90
NHS, 82
Oculus Rift, 107
Ofcom, 121

office, 61
Oracle, 110
OU, 83
pensieve, 17
Peter Thiel, 22, 102
phaser, 18
phishing, 49
PixelSense, 19
planes, 71
Pokémon Go, 40, 113
policy, 121
politics, 5, 31, 97, 114
productivity, 98
public services, 80
PWC, 88
quantum computing, 48
quantum entanglement, 48
remembrall, 17
replicator, 19, 24
restaurants, 68
retail, 67
robotics, 43, 65, 113
robots, 23, 69
rollable, 22
SAP, 110
satellite, 64, 76
Second Life, 39
security, 48, 54
Sigfox, 111
SiLabs, 111
singularity, 41, 99
Siri, 4, 41, 42, 71, 92, 113, 115
Skype, 36
smart home, 52
sneakoscope, 17
solar power, 44

spectrum, 121
Star Trek, 18
StarTac, 16
Strava, 76, 77, 107
Surrey, 121
TCP/IP, 35
telematics, 73
telemetry, 46
Tesla, 74, 104
TI, 111
time travel, 17
toilet, 56, 61, 93
trains, 72
transporter, 19
travelling, 71
Twitter, 5, 92, 107
Uber, 31, 87, 90, 91, 96, 105, 108, 109
utopia, 86
vacuum cleaners, 23, 52, 54, 55, 56, 62, 70, 95, 98, 113
video conferencing, 7
VR, 4, 22, 35, 37, 39, 58, 78, 79, 95, 107, 114
warehouses, 67
warp-drive, 18
Watson, 81
Weightless, 38, 121
Wi-Fi, 36, 52, 54, 55, 56, 58, 61, 62, 68, 71, 72, 73, 87, 93, 94, 96, 109
Wikipedia, 6
X-Box, 21
zero-hours, 87
Zoom, 64

Index

William Webb

William is CEO at Webb Search Consulting, a company specialising in providing the highest level of advice in matters associated with wireless technology and regulatory matters. He is also CEO of the Weightless SIG, the standards body developing a new global M2M technology. He was President of the IET – Europe's largest Professional Engineering body - during 14/15.

He was one of the founding directors of Neul, a company developing machine-to-machine technologies and networks, which was formed at the start of 2011 and subsequently sold to Huawei in 2014. Prior to this William was a Director at Ofcom where he managed a team providing technical advice and performing research across all areas of Ofcom's regulatory remit. He also led some of the major reviews conducted by Ofcom including the Spectrum Framework Review, the development of Spectrum Usage Rights and most recently cognitive or white space policy. Previously, William worked for a range of communications consultancies in the UK in the fields of hardware design, computer simulation, propagation modelling, spectrum management and strategy development. William also spent three years providing strategic management across Motorola's entire communications portfolio, based in Chicago.

William has published 16 books including "The 5G Myth", over 100 papers, and 18 patents. He is a Visiting Professor at Surrey and Southampton Universities, an Adjunct Professor at Trinity College Dublin, and a Fellow of the Royal Academy of Engineering, the IEEE and the IET. In 2015 he was awarded the Honorary Degree of Doctor of Science by Southampton University in

About the author

recognition of his work on wireless technologies and Honorary Doctor of Technology by Anglia Ruskin University in honour of his contribution to the engineering profession. His biography is included in multiple "Who's Who" publications around the world. William has a first class honours degree in electronics, a PhD and an MBA. He can be contacted at wwebb@theiet.org.

www.ingramcontent.com/pod-product-compliance
Lightning Source LLC
Chambersburg PA
CBHW071211240526
45470CB00018B/1706